Springer Series in Reliability Engineering

Series Editor

Professor Hoang Pham
Department of Industrial and Systems Engineering
Rutgers, The State University of New Jersey
96 Frelinghuysen Road
Piscataway, NJ 08854-8018
USA

Other titles in this series

The Universal Generating Function in Reliability Analysis and Optimization
Gregory Levitin

Warranty Management and Product Manufacture
D.N.P. Murthy and Wallace R. Blischke

Maintenance Theory of Reliability
Toshio Nakagawa

System Software Reliability
Hoang Pham

Reliability and Optimal Maintenance
Hongzhou Wang and Hoang Pham

Applied Reliability and Quality
B.S. Dhillon

Shock and Damage Models in Reliability Theory
Toshio Nakagawa

Risk Management
Terje Aven and Jan Erik Vinnem

Satisfying Safety Goals by Probabilistic Risk Assessment
Hiromitsu Kumamoto

Offshore Risk Assessment (2nd Edition)
Jan Erik Vinnem

The Maintenance Management Framework
Adolfo Crespo Márquez

Human Reliability and Error in Transportation Systems
B.S. Dhillon

Complex System Maintenance Handbook
D.N.P. Murthy and Khairy A.H. Kobbacy

Recent Advances in Reliability and Quality in Design
Hoang Pham

Product Reliability
D.N.P. Murthy, Marvin Rausand and Trond Østerås

Mining Equipment Reliability, Maintainability, and Safety
B.S. Dhillon

Advanced Reliability Models and Maintenance Policies
Toshio Nakagawa

Justifying the Dependability of Computer-based Systems
Pierre-Jacques Courtois

Reliability and Risk Issues in Large Scale Safety-critical Digital Control Systems
Poong Hyun Seong

Risks in Technological Systems
Torbjörn Thedéen and Göran Grimvall

Maintenance for Industrial Systems
Riccardo Manzini, Alberto Regattieri, Hoang Pham and Emilio Ferrari

Jinkyun Park

The Complexity of Proceduralized Tasks

Springer

Jinkyun Park, Dr.
Korea Atomic Energy Research Institute (KAERI)
Taejon 305-600
Republic of Korea
kshpjk@kaeri.re.kr

ISBN 978-1-4471-2512-9 e-ISBN 978-1-84882-791-2

DOI 10.1007/978-1-84882-791-2

Springer Series in Reliability Engineering ISSN 1614-7839

A catalogue record for this book is available from the British Library

Cover design: deblik, Berlin, Germany

Printed on acid-free paper

9 8 7 6 5 4 3 2 1

springer.com

삶은 끊임없는 즐거움의 연속임을 항상 일깨워주는 은수와 은상, 두 딸이 주는 삶의 즐거움을 같이할 수현, 그리고 변함없는 사랑이 무엇인지 알려 주신 할머니께 이 책을 바칩니다.

Preface

We think we have scientific knowledge when we know the cause.
(Aristotle, Posterior Analytics Book II, Part 11)

About 12 years ago, when I was a graduate student, many people were concerned about my Ph.D. topic – investigating the effect of the complexity of proceduralized tasks on the performance of human operators working in nuclear power plants. Although they agreed with the fact that procedures (especially emergency operating procedures) play a crucial role in securing the safety of nuclear power plants, it was amazing that most of them pointed out a very similar issue: "I cannot understand why operating personnel see any difficulty (or complexity) in conducting procedures, because all that they have to do is to follow a simple IF-THEN-ELSE rule as written." Actually, this issue is closely related to one of the main questions I was recently asked, such as "Don't you think your work is too academic to apply to actual procedures?" or "I guess we don't need to consider the complexity of procedures, because we can develop a good procedure using many practical procedure writers' guidelines. Then what is the real contribution of your work?"

I absolutely agree with the latter comment. Yes, we can develop a good procedure with the support of many practical and excellent guidelines. However, I would like to emphasize one more thing – existing guidelines seem to mainly focus on limited facets that cover some of the aspects needed to make a good procedure. For example, traditionally, most procedure writers' guidelines have recommended the use of easy sentence structures, clear writing styles, and consistent vocabularies that would be essential for specifying what should be done by operating personnel. I think these recommendations stemmed from the belief that all anticipated situations can be controlled by performing chronological actions as prescribed in a written procedure. Unfortunately, it is evident that we cannot develop a procedure that covers every situation. In addition, since procedure writers are highly experienced and possess a lot of domain-specific knowledge, they have frequently developed procedures not for less experienced people but for themselves. As a result, less experienced people have to solve their problems using a procedure that is too ambiguous or difficult to actually follow in a real-life. For this reason, from the point of view of an engineer, I started my research in order to search for a viable solution that is able to overcome this limitation.

Personally, I believe that one of the important virtues of a good engineer is the ability to provide a practical solution, such as a creative design or an outstanding idea, which actually works and has a sound technical foundation. From this standpoint, I summarized the results of my research with the associated technical solutions, which have been studied for several years, in this book. The goal of my research is to develop a systematic framework, not only by which the complexity of proceduralized tasks can be properly quantified but also from which effective countermeasures or remedial actions to reduce it can be naturally deduced. To this end, I have tried to combine a series of multidisciplinary works that seem to be closely related to the quantification of task complexities. For example, I introduced the evaluation paradigm of software complexity to provide a technical basis for quantifying the complexity of proceduralized tasks, and I adopted a classical but still valuable theory from cognitive engineering, which deals with the decision making process of human operators. In addition, I took advantage of traditional procedure writers' guidelines as well as principles in order to incorporate useful insights about the development of procedures. This implies that the readers of this book should possess basic knowledge about software engineering and cognitive engineering. In addition, since detailed examples about the quantification of proceduralized tasks are given based on a series of tasks to be performed by operating personnel working in nuclear power plants, it would be better for the readers to be familiar with nuclear engineering as well as the procedures of nuclear power plants.

This book starts with an introduction providing a motivation that ties together the three technical parts of this book: a fundamental concept (Part I), the development of a systematic framework to quantify the complexity of proceduralized tasks (Part II), and several promising applications pertaining to the developed framework (Part III). Although this book was written to be read in a linear fashion, readers may read it in many different ways. For example, those who just want to know an overview on the importance of procedures (e.g., why we need to use procedures or why we have to consider the complexity of proceduralized tasks) can read the two chapters included in Part I. If readers want to learn about a practical contribution to the evaluation of task complexities, they can read Part III, which deals with how the developed framework can be used to resolve several pending issues about the performance of human operators. In contrast, if readers would like to focus on the technical details of quantifying the complexity of proceduralized tasks, they can read the six chapters that make up Part II.

When I started to write this book, I was very nervous because many people told me that writing is a very solitary work. However, in the course of writing the book, I realized that writing was definitely not an isolated work but a kind of social endeavor through which I could enrich the contents of my book with the vast knowledge as well as diverse experiences of other people. In this regard, I deeply appreciate the encouragement of Dr. Jaejoo Ha and Dr. Joon-Eun Yang at KAERI who continually emphasized why I must write this book. In addition, the technical comments of Dr. Wondea Jung at KAERI were insightful for evolving a theoretical background on quantifying the complexity of proceduralized tasks. Dr. Dong-Han Ham of Middlesex University also provided excellent comments that were very helpful in improving the theoretical foundation of the book.

However, I would be remiss if I did not mention the sincere support of the operating personnel and training instructors working at the reference nuclear power plants. Without their help, this book would likely have turned out to be full of long-winded and hypothetical explanations lacking any useful insight. Through this book, I would like to express my heartfelt appreciation to all of them.

Jinkyun Park

Daejeon, Republic of Korea
April 2009

Integrated Safety Assessment Division
Korea Atomic Energy Research Institute

Contents

Abbreviations

ABWR	Advanced boiling light-water-cooled and moderated reactor
ACG	Action control graph
AF	Abstraction function
AGR	Advanced gas-cooled, graphite-moderated reactor
AH	Abstraction hierarchy
AHC	Abstraction hierarchy complexity
AHG	Abstraction hierarchy graph
ANOVA	Analysis of variance
ATWS	Anticipated transient without scram
BWR	Boiling light-water-cooled and moderated reactor
CBP	Computer-based procedure
CC	A feature pertaining to an action that requires a continuous control
CF	Component function
CIAS	Containment isolation actuation signal
CMP	Comprehension
CPE	Cognitive procedure engineering
CR	Character recognition
CS	Containment spray
CSAS	Containment spray actuation signal
CSF	Critical safety function
DA	Distinctive action
DBA	Design basis accident
DEG	Designated means
DI	Distinctive information
DMO	Departure from monotonic optimization
ED	Engineering decision
EDC	Engineering decision complexity
EDG	Engineering decision graph
EID	Ecological interface design
EIP	Elementary information process
EO	Electrical operator
EOP	Emergency operating procedure
ESDE	Excess steam demand event
FBR	Fast breeder reactor
GCR	Gas-cooled, graphite-moderated reactor

HMI	Human-machine interface
HPSI	High pressure safety injection
HRA	Human reliability analysis; human reliability assessment
IAEA	International Atomic Energy Agency
ICC	Intraclass correlation
INH	Inherent means
ISG	Information structure graph
KAERI	Korea Atomic Energy Research Institute
KSNP	Korean standard nuclear power plant
LER	Licensee event reports
LO	Local operation
LOAF	Loss of all feedwater
LOC	Line of code
LOCA	Loss of coolant accident
LOOP	Loss of offsite power
LWR	Light-water-cooled, graphite-moderated reactor
MCR	Main control room
MFIV	Main feed water isolation valves
MVT	Most violation-probable territory
NASA-TLX	National Aeronautics and Space Administration - task load index
NC	No criterion
NEI	Nuclear Energy Institute
NL	No limitation
NM	No means
NPP	Nuclear power plant
OBJ	Objective criterion
OBJ_C	Objective constraint
OPERA	Operator performance and reliability analysis
PBP	Paper-based procedure
PF	Process function
PHWR	Pressurized heavy-water-moderated and cooled reactor
PLCS	Pressurizer level control system
PWR	Pressurized light-water-moderated and cooled reactor
RCP	Reactor coolant pump
RCS	Reactor coolant system
RI	Reference information
RI_C	Reference information in CONSTRAINT
RO	Reactor operator
SBCS	Steam bypass control system
SBO	Station blackout
SEL	A feature pertaining to the selection of an appropriate action (i.e., equally acceptable actions)
SF	System function
SG	Steam generator
SGTR	Steam generator tube rupture
SI	Safety injection

SIAS	Safety injection actuation signal
SIC	Step information complexity
SLC	Step logic complexity
SRO	Senior reactor operator
SSC	Step size complexity
ST	Stress
SUB	Subjective criterion
SUB_C	Subjective constraint
TACOM	Task complexity (measure)
TLX	Task load index
TMI	Three Mile Island
TO	Turbine operator
TP	Task performance
TR	Task structurability
TS	Task scope
TU	Task uncertainty
USNRC	United States Nuclear Regulatory Commission
V&V	Verification and validation
WR	Word recognition
WWER	Water-cooled, water-moderated power reactor

List of Figures

List of Tables

1 Introduction

On November 28, 2006 at 4:00 pm, an airplane belonging to one of the small domestic carriers was approaching Jeju International Airport located in the largest island of the Republic of Korea. There were 69 passengers and 4 flight attendants on board. At 4:15 pm, the pilot of the airplane tried to land at the airport. At that time, the pilot recognized that there was a sudden rush of wind. Therefore, instead of a soft landing, where the main landing gear of the airplane touches down first, the pilot decided to attempt a hard landing with its nose landing gear. Unfortunately, in the course of landing, the nose landing gear broke off due to a mechanical failure. However, although the airplane skidded off the runway for a while, there were no serious injuries. As a consequence of this event, the airport was closed for about 3 h. Finally, at 7:45 pm, the airport returned to normal.

The above is the brief reconstruction of an event based on the report of an aircraft accident occurred at Jeju International Airport of the Republic of Korea (ARAIB 2006). It was a stroke of good luck that there were no serious injuries. However, what I want to emphasize from this event is that the airport restored its function within 3 h thanks to the Airplane Accident Emergency Response Manual (Article 2006). This manual was developed by the National Security Council of the Republic of Korea in 2005 to specify detailed responses with clear responsibilities regarding various kinds of emergency events that are likely to occur in an airport. Therefore, according to this manual, necessary counterplans were properly identified and then systematically carried out, such as escorting injured people to hospitals, removing the broken-down airplane from the runway, and cleaning up foreign objects (i.e., debris) from the runway, etc. Without this manual, it is evident that a huge amount of visible as well as invisible loss would have been inevitable. I think this event is a typical example illustrating *why we need a procedure.*

1.1 What Is a Procedure?

Without the loss of generality, we can define a procedure as a set of proceduralized tasks that present step-by-step instructions in the form of procedural steps composed of many actions (Inaba et al. 2004; Wagner et al. 1996). Figure 1.1 depicts the canonical structure of a procedure including proceduralized tasks, procedural steps, and the associated actions.

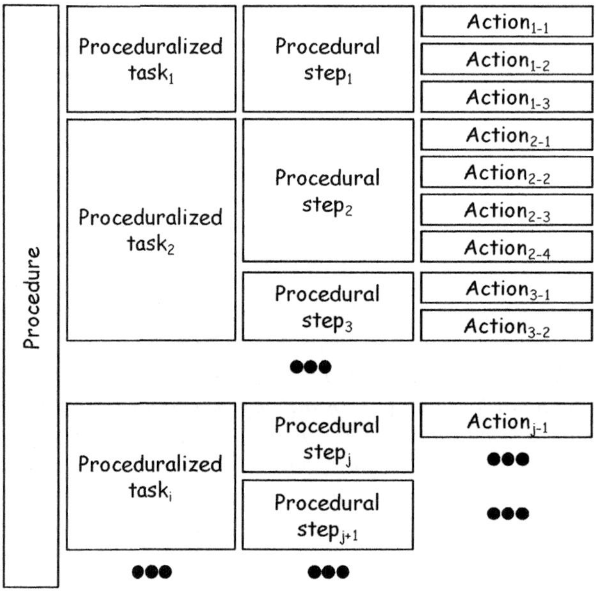

Fig. 1.1 Procedure, proceduralized tasks, procedural steps, and actions

There is no doubt that a procedure containing step-by-step instructions is very useful when people have to accomplish several specific tasks, such as safety-critical tasks, highly complex (or complicated) tasks, rarely performed tasks, and unfamiliar tasks (HSE 2007; Inaba et al. 2004; Wieringa et al. 1998). In addition, it is strongly recommended that a procedure should be developed so that even novices can follow the actions described in it, because (1) both experts and novices have shown a better performance when they used a procedure written for novices and (2) experts can be regarded as novices when they were faced with rarely performed or unfamiliar tasks (Duffy et al. 1983; Inaba et al. 2004; EPA 2001).

For these reasons, many practical principles and guidelines to develop a good procedure have been suggested for several decades. For example, Wagner et al. (1996) emphasized that "a lengthy or complicated procedure may be divided into a series of related subtasks as long as each subtask accomplishes a distinct, recognizable objective (pp. 10-48)."

It is to be noted that the same principle can be applied to proceduralized tasks and procedural steps, such as the subdivision of complicated proceduralized tasks into a series of recognizable procedural steps or the subdivision of complicated procedural steps into a series of recognizable actions (Wieringa et al. 1998). At any rate, this guideline is very important because it is anticipated that people will be able to easily identify *what should be done* or *how to do it* from procedures that consist of a series of distinct and recognizable actions.

However, it seems that a more important problem is to develop a procedure that allows people to easily and correctly carry out the proceduralized tasks in a real situation. In order to understand this issue more clearly, I would like to introduce a private episode related to baking cookies.

1.2 Recipe for a Chocolate Chip Cookie

A couple of years ago, I decided to try simple cooking, because it seemed to be a good idea to share a common memory with my daughters, Eun-su and Eun-sang. At that time, I was sure that I could make it, because not only I have general knowledge about cooking but also I know how to use kitchenware. After carefully comparing many different kinds of cuisine, I chose cookie baking because it seemed to be relatively easy. Naturally, I bought a cookbook containing many practical recipes for beginners. In the course of reviewing the contents of the cookbook, I remembered that my daughters loved chocolate chip cookies. Thus, I chose the recipe for chocolate chip cookies, which consists of (1) a list of ingredients and (2) a proceduralized task that consists of three procedural steps with the associated actions (Fig. 1.2). It is to be noted that the recipe I used was translated into English based on a recipe found on the Internet (Allrecipes 2009).

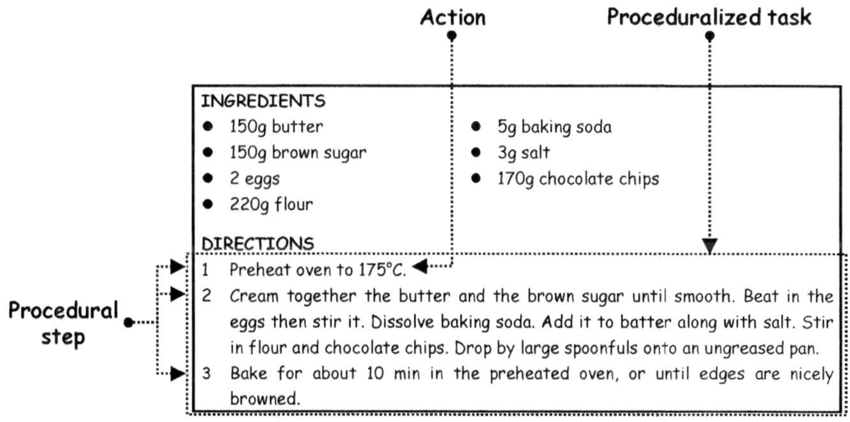

Fig. 1.2 Chocolate chip cookie recipe used by author

With this recipe, I prepared the ingredients and then preheated the oven. I put the butter in a big mixing bowl with the brown sugar and then beat it up all together with a big spoon. About 5 min later, since I thought that the batter seemed to be *sufficiently smooth*, I mixed it up again after adding the eggs, the baking soda, and the salt. Then, I mixed the batter for about 5 min with the flour and chocolate chips.

When the batter was done, I dropped it onto the ungreased pan as *large spoon-*

fuls using the big spoon that I used to mix the batter. Finally, I put the pan in the preheated oven, and waited for 10 min. But the cookies still did not seem to be done after 10 min, because the edges still remained in a *light brown color*. So, I left cookies in the oven for a few more minutes to see *browned edges*. A couple minutes later, I took the cookies out of the oven and let them cool for several hours.

I thought that I had followed the recipe exactly, but my daughters did not like my cookies. My oldest daughter, Eun-su, took a bite and said, "This cookie is too hard and has a bitter taste, dad." Moreover, Eun-sang did not even look at the cookies. It was apparent that, although I baked *edible* cookies, I failed to bake *delicious* cookies with which to impress my daughters. Thus, I explained what I did to my wife in order to find out what was the matter with my cookies. As a result, I realized that I made at least three mistakes in the course of baking the cookies.

First, although I mixed the batter for 5 min, it was not enough time to make a smooth batter with a big spoon. My wife said that I should have stirred the batter for at least 15 min and that 5 min would have been enough for a mixing machine or a hand mixer. Second, I did not sift the flour before adding it to the batter, which is a basic step in baking most cookies. Accordingly, small lumps that might cause the cookies to bake unevenly were created in the batter. Third, since the spoon I used to drop the batter was too big, the cookies came out too big. Consequently, 10 min was not enough to have nicely browned edges. This forced me to wait for several more minutes, and as a result I got hard and bitter tasting cookies. After my wife's explanation, I conceded that baking cookies was harder than it seemed.

It is to be noted that the nature of the second mistake is different from the others, because it stemmed from a lack of basic knowledge about baking cookies. Therefore, once I have gained this knowledge, I do not think I will make the same mistake again. However, it should be emphasized that the other mistakes were caused by the required actions difficult to actually carry out. That is, I felt frustration as well as confusion in performing the required actions described in the recipe, because it was quite tricky to determine such matters: *what makes for a smooth batter, how long I should mix the flour, how much makes a large spoonful, what is meant by nicely browned edge,* and so on. This strongly implies that, unless I acquire sufficient experience in baking cookies, I will probably make similar mistakes again.

1.3 What Is a Good Procedure?

In order to bake cookies, I bought a beginner's cookbook, and carefully followed the sequence of actions prescribed in the recipe. But the result was very disappointing. Fortunately, if we look at the bright side of this episode, this may serve as a nice example for elucidating a banal but always relevant issue – *what is a good procedure?*

In many cases, we are able to deduce the necessary function of a subject from the provenance of the word indicating it (i.e., etymology). As an example, let us consider the etymology of the word engineer, as depicted in Fig. 1.3 (Etymonline 2008).

Engineer

c.1325, "constructor of military engines," from O. Fr. *engigneor*, from L.L. *ingeniare* (see engine); general sense of "inventor, designer" is recorded from c.1420; civil sense, in ref. to public works, is recorded from 1606. (...)

Fig. 1.3 Etymology of "engineer"

The above etymology indicates that an engineer is an inventor or a designer who can make a machine (e.g., an engine) actually works. From this point of view, one of the necessary functions (or virtues) of an engineer is probably to provide a practical solution, such as a creative design or an outstanding idea. Consequently, we can say that a person who comes up with a practical solution is a good engineer.

In a similar vein, we are able to extract the necessary function of a good procedure from its provenance (Fig. 1.4).

Procedure

1611, "fact or manner of proceeding," from Fr. *procédure* "manner of proceeding" (1197), from O.Fr. *proceder* (see proceed). ...

Fig. 1.4 Etymology of "procedure"

From Fig. 1.4 it is evident that the word *procedure* came from *proceed*, whose the origin is shown in Fig. 1.5.

Proceed

1382, from O.Fr. *proceder* (13c.), from L. *procedere* "go forward, advance," from pro- "forward" + cedere "to go" (see cede). (...)

Fig. 1.5 Etymology of verb "proceed"

If we consider the provenance of these words simultaneously, we immediately see that one of the necessary functions of a procedure is to provide a fact (e.g., information) or manner (e.g., a detailed way of doing or the correct sequence of actions) that is helpful for going forward (i.e., carrying out) to achieve a given goal or purpose. Therefore, ideally, we can say that *a good procedure should provide crucial contents (such as information, detailed action specifications and the sequence of actions, etc.) so that people, even novices, can properly perform the required actions to achieve their goal or purpose in a real-life.*

In light of this concern, the recipe shown in Fig. 1.2 appears to be a poor procedure to some extent, because I made several mistakes in applying the recipe to baking cookies (i.e., a real-life). This problem can be understood if we compare the following three actions pertaining to one of my mistakes – *making a smooth batter*.

A1	Cream together the butter and the brown sugar *until smooth*.
A2	*Using a hand or stand mixer*, cream butter and sugars *until incorporated and smooth* (Megnut 2007).
A3	*Using a mixer fitted with paddle attachment,* cream butter and sugars together *until very light, about 5 min* (NYT 2008).

It is to be noted that, except for the first action (A1) shown in Fig. 1.2, I found the second (A2) and the third action (A3) by searching the Internet. At any rate, if we focus on the italicized parts of the three actions, we immediately realize that A2 and A3 contain more information than A1. That is, although the length of each action description in A2 and A3 is longer than in A1, A2 provides information about a useful tool to make the batter. In addition, A3 provides the operation time of the suggested tool, by which we can confirm that the batter is ready. For a beginner like me, it is assumed that a recipe containing the required actions written in the form of A3 is a good procedure, because I could have made the smooth batter more easily and correctly. This strongly implies that I could have baked more impressive cookies with a good procedure.

Here, it should be noted that I would have made the same mistakes even if I had used the new recipe shown in Fig. 1.6, which was modified based on a common principle – *the subdivision of a lengthy procedural step into a series of recognizable actions*.

DIRECTIONS

1 Preheat oven to 175°C.

2 Prepare the batter.

- 2.1 Cream together the butter and the brown sugar until smooth.
- 2.2 Beat in the eggs then stir it.
- 2.3 Dissolve baking soda.
- 2.4 Add it to batter along with salt.
- 2.5 Stir in flour and chocolate chips.
- 2.6 Drop by large spoonfuls onto ungreased pans.

3 Bake for about 10 min in the preheated oven, or until edges are nicely browned.

Fig. 1.6 Chocolate chip cookie recipe with modified second procedural step

This means that we need a novel framework that can deal with the indispensable question of how to develop a good procedure – *does a procedure contain essential instructions so that people, including novices, can perform the required actions to achieve their goal or purpose in a real situation?*

1.4 Scope of Book

Simon and Hayes (1976) pointed out that following instructions is one of the most difficult tasks encountered in daily life. Regarding this, Wright (1981) stated that there are three problems making the performance of instructions difficult. The first one is the technical correctness of procedures, because there are times when the information included in procedures is wrong. The second problem is the presentation of procedures, because the language and illustrations used in procedures are not always easily understood. The last problem is the unstructured information, because it may be inappropriately organized for the required tasks. Therefore, Wright asserted that a good procedure needs accurate information, a clear presentation, and structured information.

It should be noted that, in the previous section, we stated that a good procedure should provide essential instructions that are helpful for achieving the required tasks in a real situation. This definition is directly comparable to the last problem – *that of providing structured information*. Unfortunately, it seems that, as will be explained in Chap. 2, most research topics related to procedures seem to focus on the first and second problems issued by Wright. In this book, therefore, I would like to suggest a systematic framework for quantifying of the complexity of proceduralized tasks because it is helpful to resolve the last problem that we are concerning about.

In order to facilitate understanding the features of a quantification framework, it is helpful to provide detailed examples illustrating how to quantify the complexity of proceduralized tasks. To this end, emergency tasks prescribed in the emergency operating procedures (EOPs) of nuclear power plants (NPPs) are considered in this book. The following reasons manifest why the provision of good EOPs is critical to secure the safety of NPPs.

- *Safety-critical system* Traditionally, NPPs have actively developed diverse procedures to provide helpful instructions for most tasks to be conducted by plant personnel; one of the representative examples is EOPs (Dang et al. 1992; Mumaw et al. 1993; Wieringa and Farkas 1991). Here, as recognized from the Three Mile Island (TMI) accident, the successful performance of EOPs is a prerequisite to guarantee the safety of NPPs, because even a trivial human error could result in an irrecoverable consequence (Kemeny 1979; Wilkinson 1984).
- *Highly complicated task* NPPs are one of the most complex process control systems in the world (Perrow 1984). In addition, the operating personnel of NPPs should conduct emergency tasks prescribed in EOPs under very stressful circumstances (Kontogiannis 1996; Meister 1995). This strongly indicates that some emergency tasks could jeopardize the cognitive ability of operating personnel.

- *Rarely performed or unfamiliar task* Although the design of NPPs is very complicated, operating history has shown that the frequency of the occurrence of major accidents is very low (Amalberti 2001). However, this is a general tendency for other safety-critical systems, because considerable effort has been devoted to securing a sufficient level of safety. For example, Greenberg et al. (2005) reported that the frequency of the occurrence of major accidents in the aviation industry is 0.7×10^{-6}/h. This means that, on average, a captain should come across a major accident when he or she has flown over million hours. Accordingly, it is very natural to regard emergency tasks as rarely performed or even unfamiliar tasks.

This book consists of three parts. Part I provides some fundamental concepts that play a crucial role in quantifying the complexity of proceduralized tasks. Part II is the core of book. The six chapters included in this part will allow the reader to understand how to quantify the complexity of proceduralized tasks and to see the validity of the quantification framework. To this end, detailed explanations will be given based on the emergency tasks prescribed in the EOPs of NPPs. Then, several promising applications pertaining to the quantification framework will be reviewed in the first chapter of Part III. Finally, concluding remarks will be made in the last chapter after discussing several insights pertaining to the quantification framework.

References

Allrecipes (2009) http://allrecipes.com/Recipe/Best-Chocolate-Chip-Cookies/Detail.aspx

Amalberti R (2001) The paradoxes of almost totally safe transportation systems. Saf Sci 37:109–126

ARAIB (Aviation and Railway Accident Investigation Board) (2006). Nose landing gear collapse during landing. Aircraft Accident Report, ARAIB/AAR-0605. http://www.araib.go.kr/

Article (2006) Domestic Newspaper Article. http://news.naver.com/main/read.nhn?mode=LPOD&mid=etc&oid=078&aid=0000032408.

Dang V, Huang Y, Siu N, Carroll J (1992) Analyzing cognitive errors using a dynamic crew-simulation model. IEEE 5th Conference on Human Factors and Power Plants, pp.520–525.

Duffy TM, Curran TE, Sass D (1983) Document design for technical job tasks: an evaluation. Hum Factors 25(2):143–160

Environmental Protection Agency (2001) Guidance for preparing standard operating procedures. EPA/240/B-01/004, Washington, DC

Etymonline (2008) http://www.etymonline.com

Greenberg R, Cook SC, Harris D (2005) A civil aviation safety assessment model using a Bayesian belief network (BBN). Aeronaut J:557–568

HSE (2007) Revitalising procedures. www.hse.gov.uk/humanfactors/comah/procinfo.pdf

Inaba K, Parsons SO, Smillie R (2004) Guidelines for developing instructions. CRC, Boca Raton, FL

Kemeny JG (1979) Report of the president's commission on the accident at Three Mile Island. Washington, DC

Kontogiannis T (1996) Stress and operator decision making in coping with emergencies. Int J Hum-Comput Interact 45:75–104

Megnut (2007) http://www.megnut.com/2007/05/a-mean-chocolate-chip-cookie
Meister D (1995) Cognitive behavior of nuclear reactor operators. Int J Ind Ergonom 16:109–122
Mumaw RJ, Roth EM, Schoenfeld I (1993) Analysis of complexity in nuclear power severe accident management. In: Proceedings of the 37th Annual Meeting on Human Factors and Ergonomics, pp.377–381
NYT (2008) http://www.nytimes.com/
Perrow C (1984) Normal accident: living with high-risk technologies. Basic Books, New York
Simon HA, Hayes JR (1976) The understanding process: problem isomorphs. Cognit Psychol 8(2):165-190
Wagner D, Snyder J, Duncanson JP (1996) Human factors design guide. DOT/FAA-CT-96/1, FAA Technical Center, Washington, DC
Wieringa D, Moore C, Barnes V (1998) Procedure writing principles and practices, 2nd edn. Battelle Press, Columbus, OH
Wieringa DR, Farkas DK (1991) Procedure writing across domains: Nuclear power plant procedures and computer documentation. In: Proceedings of the 9th Annual International Conference on Systems Documentation, pp.49-58
Wilkinson CD (1984) Elements of effective control room response to emergencies. In: Lassahn PL, Majumdar D, Brockett GF (eds) Anticipated and Abnormal Plant Transients in Light Water Reactors, vol 2, Plenum, New York, pp.1049-1057
Wright P (1981) "The instructions clearly state…" Can't people read? Appl Ergonom 12:131-141

Part I
Foundation

2 Complexity of Proceduralized Tasks

As raised at the end of Sect. 1.3, it is necessary to construct a novel framework that contributes to the development of a good procedure. In order to understand this necessity more clearly, it may be helpful to review why people show a degraded performance when they are following a poor procedure in real-life.

2.1 Performing Proceduralized Tasks

Although there could be other benefits when we use a procedure, many researchers have commonly pointed out that a good procedure guarantees at least three major advantages: (1) reducing workload, (2) reducing the possibility of human error, and (3) standardizing human performance (De Carvalho 2006; Degani and Wiener 1997; Frostenson 1995; Gross 1995; HSE 2005, 2007; Roth et al. 1994). For these reasons, procedures have been widely used for many decades in large and safety-critical process control systems, such as aviation systems, railway systems, chemical/petrochemical plants and NPPs, and so on (Brito 2002; Guesnier and Heßler 1995; HSE 2007; Long 1984, Stassen et al, 1990, Wieringa et al. 1998). This indicates that a technically correct procedure is crucial to secure the safety of any human involved safety-critical system. However, in addition to the technical correctness, we need to carefully consider whether a procedure is actually able to be carried out with any undue workload. Regarding this, let us consider Fig. 2.1, which shows two examples of the allocation of cognitive resources in conducting proceduralized tasks (Wieringa et al. 1998).

In Fig. 2.1, the circle represents the total amount of available cognitive resources that people can devote to performing a proceduralized task. First, people need to devote their cognitive resources to recognizing characters they read (character recognition: CR) as well as to recognize words formed by characters (word recognition: WR). After that, they need to boil down what is to be done by understanding the meaning of a whole description formed by characters and words (comprehension: CMP). In addition, people need to devote cognitive resources to actually performing what they have to do (task performance: TP), such as remembering the location of a controller or recalling how to manipulate it, etc. However, if people have to complete proceduralized tasks in an unstable environment (or stressful circumstance, such as a severe time pressure or rapidly changing circumstance, etc.), they need to use additional cognitive resources to override the adverse effects of it (i.e., ST). A loss of concentration is a good example of the adverse effects of an unstable environment. Therefore, although the appearance of adverse effects may vary from person to person, it is frequently observed that the amount of available cognitive resources for conducting a proceduralized task is not sufficient in an unstable environment.

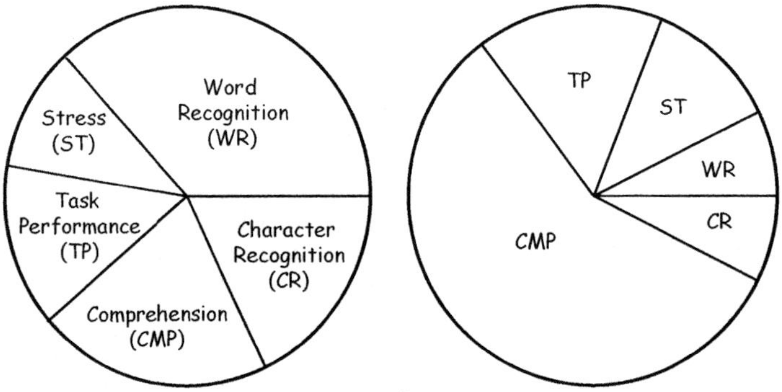

a b

Fig. 2.1 Hypothetical cognitive resource allocations related to carrying out proceduralized tasks (p. 14 of Wieringa et al. 1998)

From this concern, Fig. 2.1a shows an example of the allocation of cognitive resources when people are faced with a proceduralized task containing unfamiliar characters and words. This case may correspond to a mechanic who is trying to calculate the amount of a tax refund using a standard accounting procedure that contains many unfamiliar financial terms. In this case, it is natural to expect that the mechanic is likely to show a degraded performance (e.g., taking a long time to finish the calculation) or make a mistake (e.g., wrong calculation), because he or she will probably not be able to use a sufficient amount of cognitive resources to identify what should be done (CMP) or to carry out what he/she have to do (TP). Moreover, the effect of an unstable environment would be amplified in this case, because there are few cognitive resources to deal with it. Similarly, as shown in Fig. 2.1b, if people have to devote significant cognitive resources to CMP, they are also apt to show a degraded performance or make a mistake. Consequently, in order to avoid the degradation of human performance (or making a mistake), it is very important to develop a procedure that does not challenge the cognitive ability of people.

As a practical remedy, therefore, many procedure writers' guidelines have been developed to enhance the comprehension of proceduralized tasks (i.e., CMP) by manipulating their format, such as sentence structures, font sizes, writing styles, and vocabularies used for the description of the required actions (Brune and Weinstein 1983; EPA 2001; Fuchs et al. 1981; USNRC 1982; Wieringa et al. 1998). For example, let us reconsider two recipes shown in Fig. 1.2 and 1.6 simultaneously. From the point of view of CMP, the second procedural step in Fig. 1.2 has a problem, because it seems to be too unstructured to easily identify what should be done. In contrast, most people will easily identify what they have to do from Fig. 1.6, because a long procedural step is broken down into many distinct and recognizable actions.

It is to be noted that an enhancement of comprehension by reformatting a lengthy proceduralized task is one of the most popular techniques in procedure writers' guidelines. That is, in the beginning, most people believed that all situations could be easily controlled if a set of chronological actions included in a procedure were performed as written in a step-by-step manner. The following statement clearly shows this belief:

> In general, a procedure is a set of rules (an algorithm) which is used to control operator activity in a certain task. Thus, an operating procedure describes how actions on the plant (manipulation of control inputs) should be made if a certain system goal should be accomplished. The sequencing of actions, i.e., their ordering in time, depends on plant structure and properties, nature of the control task considered (goal) and operating constraints (Lind 1982, p. 5).

Accordingly, enhancing the comprehension of a proceduralized task has been regarded for a long time as a fundamental issue in the development of a good procedure. However, Dien (1998) pointed out that a procedure seems to be useful not as a tool for helping people to control a process but as a tool to control people. In other words, it is necessary to realize that people, especially those who are working in a large and safety-critical process control system, have to cope with a rapidly changing situation using a predefined procedure. This implies that performing a procedure is not a simple rule-following task but a problem-solving one that requires high-level cognitive activities as well as skills (Dien 1998; Grosdeva and Montmollin 1994; Kontogiannis 1999a; Roth et al. 1994; Wright and McCarthy 2003). For example, Brito (2002) says the following:

> Pilots' knowledge, expertise and know-how significantly influence the following of written procedures. These cognitive functions enable them to evaluate the situation, to categorize information presented, to evaluate the relevance and the feasibility of information presented, to plan and to execute adequate actions at the proper time (p. 242).

In addition, Spurgin et al. (1988) make the following observation:

> The procedures are very logically structured. The structure of which is related to the key process variable (symptoms) to be observed. Most accidents perturb the plant so as to affect all or a large number of key symptoms. Under these conditions the control-room crew have to simultaneously track several branches of the logic trees. This places a severe burden on the operators. They have to identify the symptoms, evaluate the symptoms that apply and interpret the procedures to carry out the recommended actions (p. 137).

Let us assume a situation in which novices are trying to bake cookies using the recipe shown in Fig. 1.6. Although novices can easily comprehend what they have to do, they may spend additional cognitive resources in the course of performing several ambiguous actions, such as deciding whether the batter is sufficiently smooth or not. That is, since this recipe forces novices to determine the condition of the batter without any specific decision criterion, they may feel a burden to perform the required action in a real situation.

In some respect, this is even a natural phenomenon, because we cannot make an *almighty* procedure describing precise actions in each and every situation. Unfortunately, this problem engenders an adverse effect – *people in a large and safety-critical process control system need to devote cognitive resources not only to identify what they have to do but also to properly conduct it.* For example, in an extreme case, the allocation of cognitive resources could be like Fig. 2.2.

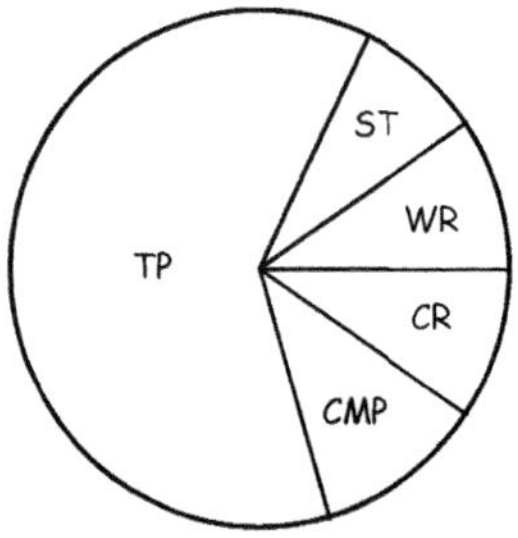

Fig. 2.2 Example of the allocation of cognitive resources when the performance of a proceduralized task is extremely complicated

Obviously, this one-sided allocation is very vulnerable to the degradation of human performance as well as human error, because there are few cognitive resources to conduct the other activities (i.e., CR, WR, CMP and ST). Nevertheless, as mentioned before, it is surprising that most procedure writers' guidelines have mainly focused on the enhancement of a procedure by managing CR, WR and CMP. For this reason, I think that it is critical to develop a systematic framework by which the quality of procedures can be evaluated from the point of view of TP. One promising way to resolve this problem is to measure the complexity of proceduralized tasks, because it is expected that the more the complexity increases, the more the demand of cognitive resources increases.

2.2 Managing the Complexity of Proceduralized Tasks

Related studies have revealed that the amount of effort to be put into a cognitive task (e.g., choice or selection) could be measured as the sum of well-defined units of thought (or elementary information process, EIP), such as *READ*, *RETERIVE*, *MOVE*, *ADD*, etc. (Campbell and Gingrich 1986; Jiang and Klein 2000; Johnson and Payne 1985; Shugan 1980; Sintchenko and Coiera 2002). With this result, if we define an effort as *the total use of cognitive resources required to complete a task* (Russo and Dosher 1983), then it is expected that the amount of effort will be proportional to the complexity of proceduralized tasks. For example, Campbell and Gingrich (1986) articulated that a complicated task places substantial cognitive demands on a task-doer for comprehension (i.e., CMP) and execution (i.e., TP). This strongly indicates that people have to spend more cognitive resources in the course of carrying out a complicated proceduralized task because they need to process more cognitive activities compared to an easy one (Arend et al. 2003; Jonassen 2000). Accordingly, Fig. 2.3 clarifies why we have to manage the complexity of proceduralized tasks.

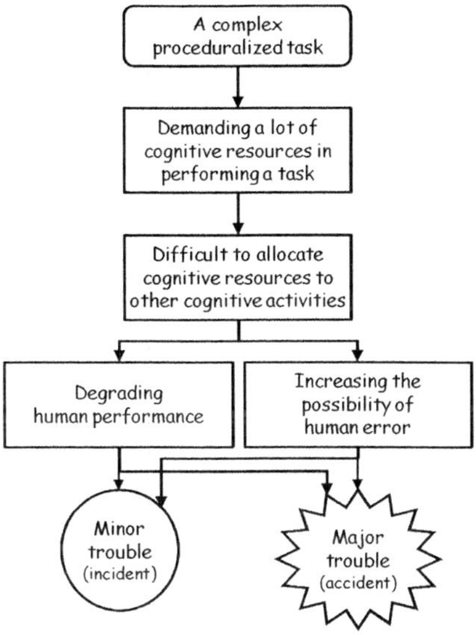

Fig. 2.3 The effect of a complicated proceduralized task on unfavorable consequences

Above all, complicated proceduralized tasks may compel people to spend additional cognitive resources on TP. This results in a decrease in cognitive resources to be spent on other cognitive activities, such as CR, WR or CMP. Because of the lack of cognitive resources, people are likely to either show a degraded performance or make a mistake (Morris and Rouse 1985; Rouse and Rouse 1983; Woods 1990; Woods et al. 1990). In most cases, a degraded performance and human error just cause minor troubles or incidents with a tolerable consequence. However, there are times when an impaired performance as well as human error are unacceptable because they trigger irreversible consequences. For example, a deviation from procedures is one of the typical human errors that culminate in major troubles or accidents in a large and safety-critical process control system (Degani and Wiener 1990, 1997; Lauber 1989; Marsden 1996). Here, it should be noted that a large portion of these deviations is due to the complexity of proceduralized tasks. That is, since people frequently feel an excessive workload due to a complex procedure, they are susceptible to unintended deviations from it. Degani and Wiener (1990) referred to this deviation as *distraction-due-to-workload* (p. 33). A more interesting point is that the complexity of proceduralized tasks seems to contribute to the occurrence of violations (Gross 1995; Hale 1990; Wood 1986).

In general, a violation implies any intended deviation from rules, procedures, or regulations (HSE 1995; Reason et al. 1998). Nevertheless, most violations can be regarded as not malicious actions (e.g., *sabotage*) but a kind of optimized response to satisfactorily perform the required tasks under a given constraint (Gross 1995; Helmreich 2000; HSE 1995; Reason et al. 1998). For example, Dien (1998) stated that "The operators are often called on to respond to situations or events that are not explicitly featured in the procedure. … Some actions required by the procedure may not be totally clear, thereby obliging the operators to take real-time initiatives and decisions in order to overcome any ambiguity (p. 183)." Therefore, as Degani and Wiener (1990) commented, it is meaningful to regard violations as "Deviations from those practices deemed necessary to maintain the safe operations of a hazardous system (p. 42)."

Ironically, operating records have clearly shown that violations are one of the primary sources of major accidents (Perrow 1984; Wiegmann and Shappell 2001). Therefore, from the point of view of securing a sufficient level of safety, it is very important to understand why people violate a procedure. In this regard, several researchers have provided insightful clues. Degani and Wiener (1997) stated that "A procedure that is ponderous and is perceived as increasing workload, and/or interrupting smooth flow of cockpit tasks, will probably be ignored (p. 306)." In addition, Marsden (1996) pointed out that "The operators reported that working with procedures made work much less rewarding and the job more difficult than it would otherwise be (p. 111)." Finally, Macwan and Mosley (1994) have the following to say:

> It is assumed that all plant personnel act in a manner they believe to be in the best interests of the plant. Any intentional deviation from standard operating procedures is made because the employee believes their method of operation to be safer, more economical, or more efficient or because they believe performance as stated in the procedure to be unnecessary (Macwan and Mosley 1994, p. 143).

The above statements emphasize one common tendency as depicted in Fig. 2.4.

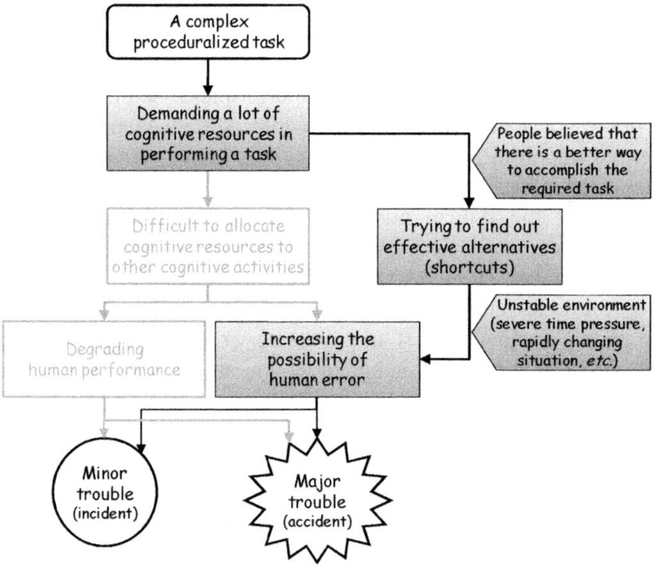

Fig. 2.4 Side effect of a complicated proceduralized task – searching for shortcuts

That is, although there would be many other reasons for violations, people are likely to deviate from a procedure if they believe that there is a better way to accomplish a complicated proceduralized task (i.e., saving cognitive resources by customizing the complicated proceduralized task). It is very fortune that, in most cases, the result of these violations is not harmful but even effective to a certain extent. However, if a less harmful violation is combined with an unstable environment, it is strongly expected that the possibility of human error will drastically increase (Williams 1988; Reason et al. 1998). This means that we have to carefully consider the side effect of a complicated proceduralized task.

Consequently, as illustrated in Fig. 2.5, there is no doubt that we have to actively manage the complexity of proceduralized tasks from the point of view of TP. Otherwise, we would probably face a difficulty in reducing the possibility of major troubles or accidents triggered by complicated proceduralized tasks.

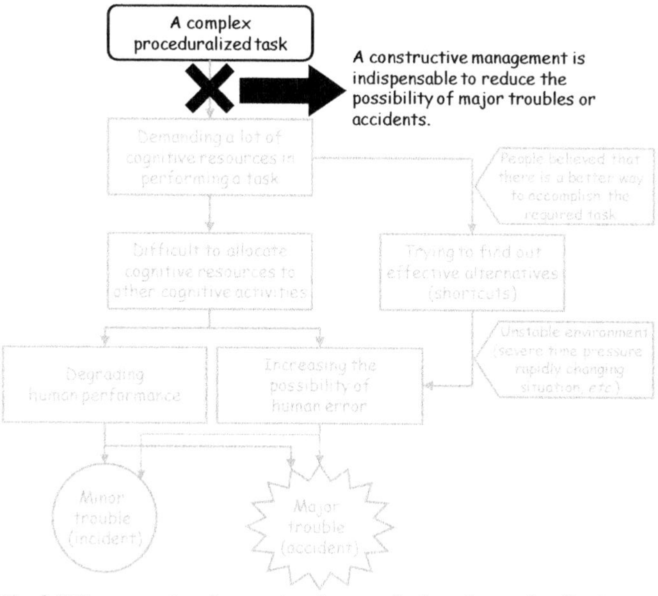

Fig. 2.5 The necessity of managing the complexity of proceduralized tasks

References

Arend I, Colom R, Botella J, Contreras MJ, Rubio V, Santacreu J (2003) Quantifying cognitive complexity: evidence from a reasoning task. Personal Individ Differences 35:659–669

Brito G (2002) Towards a model for the study of written procedure following in dynamic environments. Reliabil Eng Syst Saf 75:233–244

Brune RL, Weinstein M (1983) Checklist for evaluating emergency operating procedure used in nuclear power plants. NUREG/CR-2005, Washington, DC

Campbell DJ, Gingrich KF (1986) The interactive effects of task complexity and participation on task performance: a field experiment. Organizat Behav Hum Decis Processes 38:162–180

De Carvalho PVR (2006) Ergonomic field studies in a nuclear power plant control room. Prog Nuclear Energy 48:51–69
Degani A, Wiener EL (1990) Human factors of flight-deck checklists: the normal checklist. NASA/CR-177549
Degani A, Wiener EL (1997) Procedures in complex systems: the airline cockpit. IEEE Trans Syst Man Cybern 27(3):302–312
Dien Y (1998) Safety and application of procedures, or 'how do they have to use operating procedures in nuclear power plants?' Saf Sci 29:179–187
Environmental Protection Agency (2001) Guidance for preparing standard operating procedures. EPA/240/B-01/004, Washington, DC
Frostenson CK (1995) Lessons learned from occurrences involving procedures at LOS ALAMOS National Laboratory in 1994. In: Proceedings Human Factors and Ergonomics Society (HFES) Annual Meeting, 39:1033–1037
Fuchs F, Engelschall J, Imlay G (1981) Evaluation of emergency operating procedures for nuclear power plants. NUREG/CR-1875, Washington, DC
Grosdeva T, Montmollin M (1994) Reasoning and knowledge of nuclear power plant operators in case of accidents. Appl Ergonom 25(5):305–309
Gross RL (1995) Studies suggest methods for optimizing checklist design and crew performance Flight Saf Dig 14(5):1–10
Guesnier G, Heßler C (1995) Milestones in screen-based process control. Kerntechnic 60(5/6):225–231
Hale AR (1990) Safety rules O.K? J Occupat Accid 12:3–20
HSE (1995) Improving compliance with safety procedures reducing industrial violations. http://www.hse.gov.uk/humanfactors/comah/improvecompliance.pdf
HSE (2005) Inspection toolkit – human factors in the management of major accident hazards. www.hse.gov.uk/humanfactors/comah/toolkitintro.pdf
HSE (2007) Revitalising procedures. www.hse.gov.uk/humanfactors/comah/procinfo.pdf
Helmreich RL (2000) On error management: lessons from aviation. British Med J 320:781–785
Jiang JJ, Klein G (2000) Side effects of decision guidance in decision support systems. Interact Comput 12:469–481
Johnson EJ, Payne JW (1985) Effort and accuracy in choice. Manage Sci 31:395–414
Jonassen DH (2000) Toward a design theory of problem solving. Educat Technol Res Develop 48(4):63–85
Kontogiannis T (1999a) Applying information technology to the presentation of emergency operating procedures: implication for usability criteria. Behav Inf Technol 18(4):261–276
Lauber JK (1989) NORTHWEST 255 at DTW: anatomy of a human error accident. Hum Factors Aviat Med 30(4):1–8
Lind M (1982) The use of flow models for design of plant operating procedures. RISØ-M-2341, Risø
Long AB (1984) Computerized operator decision aids. Nuclear Saf 25(4):512–524
Macwan A, Mosleh A (1994) A methodology for modeling operator errors of commission in probabilistic risk assessment. Reliabil Eng Syst Saf 45:139–157
Marsden P (1996) Procedures in the nuclear industry. In: Stanton N (ed) Human Factors in Nuclear Safety. Taylor & Francis, London
Morris NM, Rouse WB (1985) Review and evaluation of empirical research in troubleshooting. Hum Factors 27(5):503–530
Perrow C (1984) Normal accident: living with high-risk technologies. Basic Books, New York
Reason J, Parker D, Lawton R (1998) Organizational controls and safety: the varieties of rule-related behavior. J Occupat Organizat Psychol 71:289–304
Roth EM, Mumaw RJ, Lewis PM (1994) An empirical investigation of operator performance in cognitively demanding simulated emergencies. NUREG/CR-6208, Washington, DC
Rouse WB, Rouse SH (1983) Analysis and classification of human error. IEEE Trans Syst Man Cybern SMC-13(4):539–549

Russo JE, Dosher B (1983) Strategies for multiattribute binary choice. J Exp Psychol: Learn, Mem Cognit 9:676–696

Shugan SM (1980) The cost of thinking. J Consumer Res 7(2):99–111

Sintchenko V, Coiera E (2002) Which clinical decision benefit from automation? a task complexity approach. In: Surjan G, Engelbrecht R, McNair P (eds) Proceeding of MIE2002:639–648, IOS Press, Amsterdam

Spurgin AJ, Orvis DD, Cain DG, Yau CC (1988) Testing an expert system: Testing the emergency operating procedures tracking system. In: Proceedings of the IEEE 4th Conference on Human Factors and Power Plants, Monterey, CA, pp.137–140

Stassen HG, Johannsen G, Moray N (1990) Internal representation, internal model,human performance model and mental workload. Automatica 26(4):811–820

USNRC (1982) Guidelines for the preparation of emergency operating procedures. NUREG-0899, Washington, DC

Wiegmann DA, Shappell SA (2001) A human error analysis of commercial aviation accidents using human factors analysis and classification system (HFACS). DOT/FAA/AM-01/3, Washington, DC

Wieringa D, Moore C, Barnes V (1998) Procedure Writing Principles and Practices, 2nd edn. Battelle Press, Columbus, OH

Williams JC (1988) A data-based method for assessing and reducing human error to improve operational performance. In: Proceedings of the IEEE 4th Conference on Human Factors in Power Plants, Monterey, CA, pp.436–450

Wood RE (1986) Task complexity: definition of the construct. Organizat Behav Hum Decis Processes 37:60–82

Woods DD (1990) On taking human performance seriously in risk analysis: comments on Dougherty. Reliabil Eng Syst Saf 29:375–381

Woods DD, Roth EM, Pople, HE Jr. (1990) Modeling operator performance in emergencies. In: Proceedings on the 34th Human Factors and Ergonomics Society Annual Meeting, Orlando, FL, pp.1132–1136

Wright P, McCarthy J (2003) Analysis of procedure following as concerned work. In: Hollnagel E (ed) Handbook of Cognitive Task Design, Lawrence Erlbaum Associates, London, pp. 679–701

3 Significant Complexity Factors

As shown in the previous chapter, we have to develop a novel framework to evaluate the complexity of proceduralized tasks. To this end, it is natural to start by identifying what factors make the performance of proceduralized tasks complicated. In other words, instead of many complexity factors pertaining to CR, WR, and CMP, it is necessary to consider different factors that could annoy people by demanding additional cognitive resources for TP. For this reason, many works dealing with causal factors regarding the complexity of a proceduralized task have been reviewed. As a result, a total of nine categories of complexity factors were ddistinguished. In this chapter, I would like to explain the meaning as well as the characteristics of each category.

It is to be noted that, in the course of this literature survey, two basic principles have been applied to the selection of complexity factors. Therefore, before explaining task complexity factors, it is helpful to clarify the basic principles that I have adopted.

3.1 Complexity Factors of a Process Control Task

The first principle was that, although this may sound natural, we must focus on complexity factors pertaining to the nature of the tasks being considered. Jonassen (2000) stated that "These cognitive demands are situationally specific. Arguing a case in court, for instance, would demand a different set of cognitive skills from those needed for air traffic controlling (p. 79)."

This means that, before identifying significant factors that make the performance of proceuralized tasks complicated, we should make explicit the task type we are concerned with. In this regard, our interest is in managing the complexity of proceduralized tasks used in a large and safety-critical process control system. Therefore, we have to concentrate on complexity factors related to a process control task.

At the same time, we need to concentrate on complexity factors pertaining to a supervisory control task. For example, Stassen et al. (1990) and Johannsen et al. (1994) specified that a supervisory control generally consists of several subtasks, such as monitoring, interpreting, planning, fault management, and intervention, etc. Here, it is very interesting to consider the definition of Leitch and Gallanti (1992), who articulated that a process control task must deal with a dynamic physical system evolving over time, which consists of five primitive behaviors as listed in Table 3.1.

Table 3.1 Five primitive behaviors related to a process control task

Behavior	Meaning
Decision	An action generating hypotheses or conclusions that satisfy given constraints or specifications
Prediction	An action generating future states from the present state using an implicit or explicit model of a system
Identification	An action related to determining unknown or unmeasurable states from known or assumed states
Interpretation	An action generating a situational description from observable data
Execution	An action related to the actuation of a target system

It is to be emphasized that process control tasks and supervisory control tasks resemble each other, because primitive behaviors related to process control tasks seem to be directly comparable to those of supervisory control tasks. For example, an intervention behavior is congruent with an execution behavior shown in Table 3.1, and a fault management behavior seems to be comparable with an identification behavior, etc. Accordingly, we need to search the complexity factors that would be related to either process control or supervisory control tasks.

3.2 Complexity Factors of a Novice

The second principle was that all kinds of complexity factors should be applicable not to a user's manual but to a procedure that provides practical contents for performing a process control task. This principle is closely connected to the definition of a good procedure – *the procedure should be developed so that even a novice can properly follow it.* Here, it should be noted that the novice in this book implies a human operator who already has a certain level of domain knowledge. Let us consider the following examples.

- Starts a computer using the power button.
- Starts a computer using the power button. It is located on the front panel of the computer. It is round and about the size of a quarter. You can boot the computer by pushing this button.

Here, it is presumed that the first instruction may be unclear for a person seeing a computer for the first time. This is because he or she may have felt frustration in starting the computer due to a lack of basic knowledge regarding, for example, the power button, its location and so on. In contrast, since the second instruction contains very detailed information, it is expected that even someone seeing a computer for the first time would easily boot it up.

Unfortunately, the second instruction is closer to what you would find in a user's manual rather than a procedure. This is because it simultaneously provides two kinds of descriptions that have a different purpose, such as (1) the description of an action to be done (i.e., *starts a computer using a power button*) and (2) additional descriptions about the physical form of the power button. In other words, since we are looking for complexity factors making the performance of proceduralized tasks complicated, we have to pick out those that are meaningful for a novice who has a minimum level of domain knowledge. A person with general knowledge of cooking as well as how to deal with kitchenware is a good example of a novice who is ready to use recipes (i.e., procedures). Similarly, operating personnel of NPPs who have just passed a basic training course are novices who can follow a procedure. Therefore, several factors pertaining to a lack of domain-specific knowledge, such as experience (Thelwell 1994; Maynard and Hakel 1997; Van Eekhout and Rouse 1981; Morris and Rouse 1985) or job training/skill (Li and Wieringa 2000; Leplat 1998), have been excluded from the consideration of task complexity factors. For the sake of convenience, henceforth, a *qualified operator* will be referred to as a person who is ready to follow a procedure, while an *unqualified operator* will refer to an ordinary person without a minimum level of domain knowledge.

3.3 Identifying Complexity Factors

With the aforementioned principles in mind, existing works that deal with many kinds of complexity factors have been reviewed. As a result, in total nine categories of complexity factors were identified as epitomized in Table 3.2. Appendix A summarizes all the complexity factors belonging to each category. It is to be noted that the meanings of four categories (i.e., time pressure, temporal characteristics, system characteristics, and personal characteristics) are self-explanatory from the summary in Appendix A. Therefore, more detailed explanations will be provided for the remaining categories.

3.3.1 Amount of Information and Number of Actions

The first category is the amount of information to be processed by a qualified operator. For example, it seems to be clear that a proceduralized task pertaining to operating a huge chemical plant is more complicated than that of a small domestic factory, since qualified operators have to manage more information including process alarms or process parameters, etc. Therefore, it is strongly expected that qualified operators working in the former need to spend more cognitive resources compared to those working in the latter.

Table 3.2 Categories of complexity factors

No.	Categories	Description
1	Amount of information	Amount of information to be processed by a qualified operator
2	Number of actions	Number of actions to be conducted by a qualified operator
3	Logical entanglement	Logical complexity due to the sequence of actions to be followed by a qualified operator
4	Amount of domain knowledge	Amount of domain knowledge to be considered by a qualified operator
5	Level of engineering decision	Amount of cognitive resources to be used by a qualified operator, which is needed to establish an appropriate decision criterion
6	Time pressure	Time allowed for the performance of a task
7	Temporal characteristics	Degree of a task arrival, task frequency, task overlap, etc.
8	System characteristics	Dynamic characteristics of a task due to the nature of the system
9	Personal characteristics	Aptitude, intelligence, ability, and cognitive style of a qualified operator

Similarly, the number of actions to be conducted by qualified operators is an obvious factor making the performance of proceduralized tasks complicated, because they need to use cognitive resources to properly conduct each and every action. However, this factor seems to be somewhat *superficial*, because the complexity of the cookie recipe, which includes eight actions (Fig. 1.6), is definitely different from that of an arbitrary proceduralized task that consists of two procedural steps with the same number of actions (Fig. 3.1). It is to be noted that Fig. 3.2 depicts a target system to be managed by the proceduralized task shown in Fig. 3.1.

As depicted in Fig. 3.2, there are four valves contributing to the change of the water level of a reservoir (i.e., Tank 1). First, an influx into Tank 1 is governed by IV 1, which has only two operable states – open and close. Meanwhile, CV 1 regulates the rate of an outflow from Tank 1 by continuously adjusting its open position from 0% to 100%. In addition, in order to prevent the overfill of Tank 1, there are two bypass vales (BV 1 and BV 2), which are normally in a closed state. That is, when the water level is too high, these valves can be used to provide another flow path draining the water from Tank 1. In this regard, three more categories – logical entanglement, the amount of domain knowledge, and the level of an engineering decision – are needed to reflect the *hidden* aspect of the complexity of proceduralized tasks.

RESPONSE TO THE HIGH WATER LEVEL OF TANK 1

1 IF the water level of Tank 1 is within 50~70%,
THEN
- 1.1 Close IV 1*.
- 1.2 Increase the opening position of CV 1* to 10% higher than the current position.

2 IF the water level of Tank 1 is over 70%,
THEN perform one of the following:
- 2.1 Increase outflow.
  - 2.1.1 Close IV 1.
  - 2.1.2 Increase the opening position of CV 2 to 30% higher than the current position.

OR

- 2.2 Provide bypass line.
  - 2.2.1 Open BV 1*.
  - 2.2.2 Open BV 2.

* IV, CV, and BV stand for isolation valve, control valve, and bypass valve, respectively.

Fig. 3.1 Arbitrary proceduralized task pertaining to controlling the water level of a reservoir

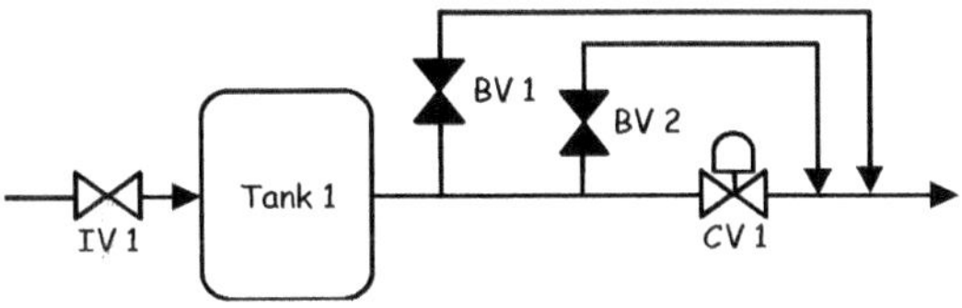

Fig. 3.2 An arbitrary system including four valves and a reservoir

3.3.2 Logical Entanglement

First, we need to consider the logical entanglement caused by the relationship of the required actions. For example, Kieras and Polson (1985) regarded the number of execution sequences needed to achieve a goal as one of the dominant complexity factors. In addition, similar comments were made by many researchers such as Leplat (1998), Li and Wieringa (2000), Sundstrom (1993), Thelwell (1994), Wood (1986), and Wood and Locke (1990). These comments can be conceptualized as *path-goal multiplicity* (Jacko and Salvendy 1996) or *multiple path-goal connections* (Campbell 1988), which indicates the number of different ways to perform a task. To explain this concept, let us consider Fig. 3.3, which illustrates the sequence of actions in the recipe shown in Fig. 1.6.

As shown in Fig. 3.3, the recipe just provides a single way to bake a cookie. If qualified operators conduct a proceduralized task like this, then they perhaps do not need to use additional cognitive resources to clarify *if they are correctly following the sequence of actions, if they are doing what they need to be done,* etc. In contrast, let us assume that qualified operators have to follow the sequence of actions depicted in Fig. 3.4, which is related to coping with the high water level of Tank 1.

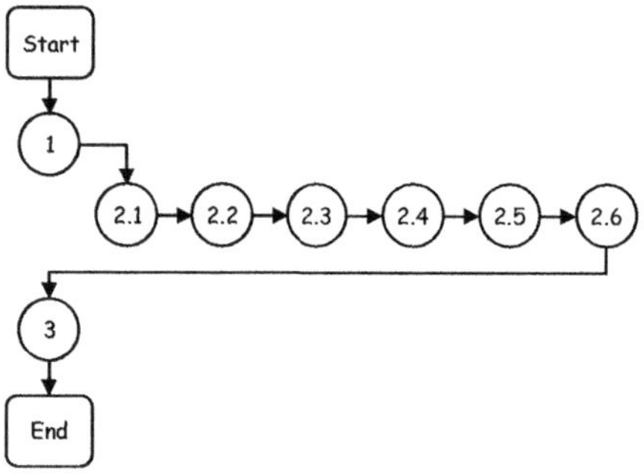

Fig. 3.3 Sequence of actions to bake chocolate chip cookies

In Fig. 3.4, there are four paths to accomplish this task. First, if the water level of Tank 1 is maintained between 50 and 70%, then qualified operators need to perform two actions (*1.1* and *1.2* in Fig. 3.4). Meanwhile, if the water level of Tank 1 is greater than 70%, then qualified operators have to select either the second or the third path to decrease the water level of Tank 1. The second path consists of two actions designed to increase the rate of an outflow from Tank 1 by opening CV 1. The third path also consists of two actions but seems to be more aggressive, because its intention is to provide additional drain channels by opening two bypass valves, BV 1 and BV 2. The last path is somewhat trivial because it says there is nothing to do if the water level is less than 50%.

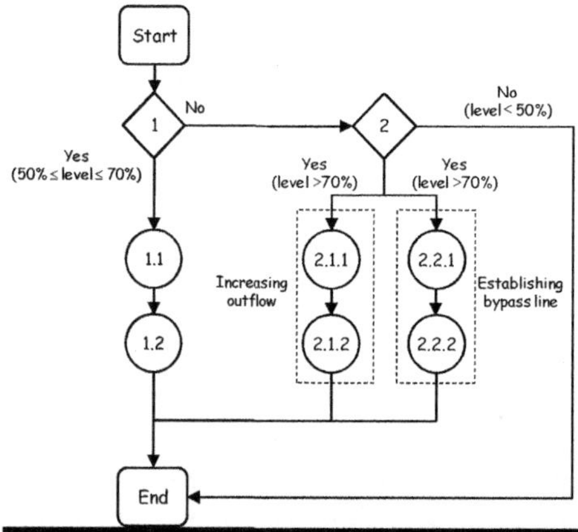

Designation	Action description
1	Determine the water level of Tank 1 is 50~70%
1.1	Close IV 1
1.2	Increase the opening position of CV 1 to 10% higher than the current position
2	Determine the water level of Tank 1 is over 70%
2.1.1	Close IV 1
2.1.2	Increase the opening position of CV 2 to 30% higher than the current position
2.2.1	Open BV 1
2.2.2	Open BV 2

Fig. 3.4 Sequence of required actions related to the proceduralized task shown in Fig. 3.1

Consequently, compared to baking a cookie, it seems that qualified operators may use additional cognitive resources to complete this task because they need to pay attention to following the correct sequence of actions with respect to the situation at hand. In general, therefore, it is expected that the greater the number of possible paths to accomplish a proceduralized task, the more cognitive resources will have to be used by qualified operators.

3.3.3 Amount of Domain Knowledge

The next category is the amount of domain knowledge, because it is natural to assume that the amount of domain knowledge for carrying out each action is not equal. Actually, the results of existing studies support this assumption, because they have revealed that qualified operators need to use their knowledge of a system in order to carry out a procedure (Boy and Brito 2000; Spangler and Peters 2001; Wright et al. 1998). In this light, it is very interesting to compare the original proceduralized task shown in Fig. 3.1 with a slightly modified proceduralized task as illustrated in Fig. 3.5.

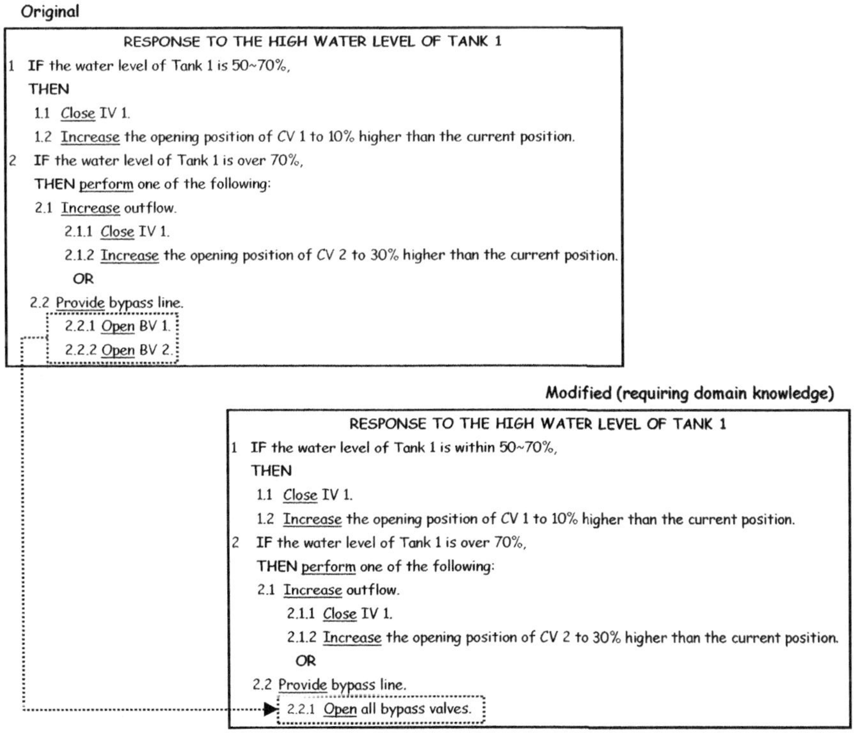

Fig. 3.5 Actions requiring different levels of domain knowledge

First, as mentioned earlier, the purpose of the required actions that are enclosed by dotted lines in the *original* task is to reduce the water level of Tank 1 by opening bypass valves. To this end, each action showed a dedicated component (i.e., BV 1 and BV 2) as a target to be manipulated by qualified operators. This means that qualified operators probably do not need to use additional cognitive resources to recall (or extract) appropriate domain knowledge, such as a component's configuration because each action is limited to component itself.

In contrast, the required action enclosed by dotted lines in the *modified* task may demand a higher level of domain knowledge because it is related not to a dedicated component but to a set of components grouped to accomplish a desired function. In other words, although the intention of this action is identical to that of two previous actions (i.e., *open BV 1* and *open BV 2*), it probably forces qualified operators to recall a kind of domain knowledge about the configuration of bypass valves, such as *how many bypass valves are there?* In addition, since a complicated process control system will include many components with many different functions, it is generally expected that the greater the number of components, the more domain knowledge will have to possess qualified operators. This strongly implies that the amount of cognitive resources needed to recall the proper domain knowledge will increase in proportion to the number of components included in the process control system being considered.

It should be noted that many researchers have reported a similar concern. For example, Rouse (1978) and Rouse and Rouse (1979) pointed to the number of components as one of the major contributors to task complexity. In addition, Allen et al. (1996), Morris and Rouse (1985), Leplat (1998), and Liao and Palvia (2000) commonly distinguished two kinds of task complexity factors, such as the number of functional relations among components as well as equipment. Moreover, although experimental data have been collected from unqualified operators, Park et al. (2008) observed that their performance seemed to be significantly affected by the amount of domain knowledge they possessed. Consequently, it is reasonable to regard the amount of domain knowledge as one of the dominant factors on the complexity of proceduralized tasks. A more detailed explanation can be found in Sect. 6.3.2.

3.3.4 Level of an Engineering Decision

Another complexity category is the level of an engineering decision, which is related to the amount of cognitive resources used to establish appropriate decision criteria for performing required actions. In order to understand the nature of an engineering decision, it may be necessary to answer two crucial questions: (1) *what is the engineering decision*? and (2) *why do qualified operators need to decide something while they are performing a proceduralized task*?

First, let us consider the following explanations given by Turk (2001) and Ditlevsen (2003), who identified an important feature of engineering decisions, respectively.

> Engineering is based on sound principles of mathematics and physics, however, not every engineering decision is based on calculations and models. Engineers also use intuition, common sense and insight when they design. The origin of such 'feelings' (i.e. intuition, common sense, etc.) could be numerous, perhaps from experiences" (see p. 247 of Turk 2001).
> In engineering decisions the usual situation is that it is generally not possible to choose the safe lottery, i.e. the lottery that for sure gives the benefit and never the loss. This can be expressed by saying that among all the possible lotteries of relevance in the considered technical problem only some of the lotteries are realizable. To be able to choose among the realizable lotteries in a rational way the decision maker must, at least partly, put the lotteries in some priority order of preference that points at a most preferred realizable lottery (see p. 167 of Ditlevsen 2003).

The above excerpts state that engineers will use not only domain-specific knowledge but also all kinds of available knowledge (such as feeling, intuition, or common sense, etc.) in order to find a practical solution to an actual problem. In a similar vein, in order to correctly perform what they have to do, qualified operators will do their best to establish proper decision criteria by using all kinds of available knowledge. It is to be noted that, for this reason, the term *the level of an engineering decision* was adopted in this book instead of *the level of a decision.*

Second, in order to explain why qualified operators need to make an engineering decision, let us recall the cookie baking episode in Chap. 1. In this episode, although I followed all the required actions very well, I made several mistakes in the course of baking cookies. This is because I failed to establish correct decision criteria, such as *the nature of a smooth batter*, *what a nicely browned edge* and so on. As a result, I got hard and bitter tasting cookies. This clearly shows that establishing proper decision criteria is crucial for accomplishing required actions.

Actually, Sundstrom (1993) pointed out that qualified operators who are working in a dynamic task environment should constantly update their perception in order to make two kinds of decisions regarding (1) what control tasks need to be accomplished and (2) how they need to be prioritized. Subsequently, Sundstrom identified several task complexity factors including (1) interrelatedness of assessment, choice, and evaluation rules, (2) interconnectedness of operational states, (3) relation between indicators and operational states, (4) the number of assessments, choices and evaluation rules, and (5) the number and relationship between conditions for assessments, choices and evaluation rules. In addition, Kieras and Polson (1985), Schmuck and Gundlach (1989), Svensson et al. (1997), and Thelwell (1994) identified similar complexity factors.

For example, let us consider the sequence of actions illustrated in Fig. 3.4, which has an unusual decision point. That is, qualified operators have to select one of the action sequences, either *increasing outflow* or *establishing a bypass line*, which is more appropriate for decreasing the water level of Tank 1. To this end, qualified operators need a decision criterion by which the proper action sequence can be determined. Unfortunately, settling on a decision criterion is harder than it seems, because qualified operators need to integrate at least two kinds of information (i.e., the trend of the water level of Tank 1 and the open position of CV 1) to assess an ongoing situation. Figure 3.6 will be helpful to illustrate this intricacy.

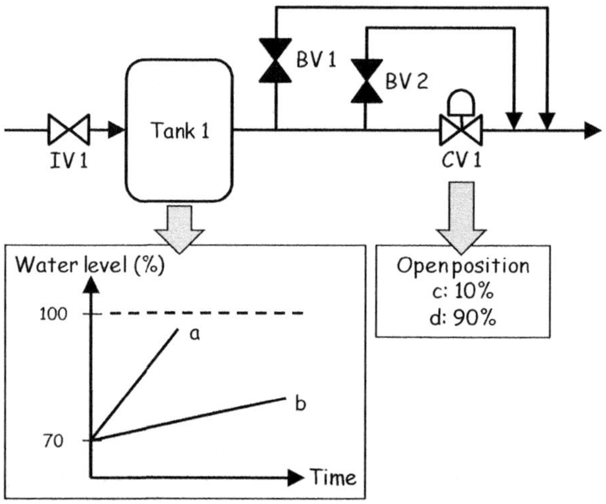

Fig. 3.6 Hypothetical situations with which qualified operators may be faced

The first situation with which qualified operators may be faced is the combination of {a, d}. This situation represents *the water level is drastically increasing* and *the open position of CV 1 is 90%*. In this situation, most qualified operators might select the action sequence related to *establishing a bypass line*. That is, since CV 1 is already opened to 90%, it is anticipated that this valve will not be able to reduce the water level that will apparently soon reach 100%. In contrast, in the situation of {b, c}, most qualified operators would probably select the action sequence pertaining to *increasing outflow*, because the gradual increment of the water level seems to be successfully compensated by increasing the open position of CV 1.

If qualified operators have to establish an appropriate decision criterion by integrating several kinds of information, it is evident that they may use a considerable amount of cognitive resources. In other words, although there is a difference in the level of depth, determining an appropriate decision criterion can be accomplished by performing a set of high-level cognitive activities (such as identification, interpretation, decision, etc.) that belong to the primitive behaviors of a process control task (Table 3.1). In addition, it is assumed that the amount of cognitive resources demanded from these cognitive activities can be explained by a series of well-defined units of thought (Campbell and Gingrich 1986; Jiang and Klein 2000; Johnson and Payne 1985; Shugan 1980; Sintchenko and Coiera 2002). This assumption seems to be empirically supported, although experimental data have been collected from unqualified operators, because Park et al. (2008) observed that their performance seems to vary with respect to the level of engineering decision. Consequently, we can say that the level of an engineering decision is one of the significant factors complicating the performance of proceduralized tasks. A more detailed explanation about the level of an engineering decision is given in Sect. 6.3.3.

3.4 Where Is the Starting Point?

Till now, the nine categories of complexity factors have been discussed from the point of view of a process control task. Roughly speaking, these categories can be regrouped as depicted in Fig. 3.7.

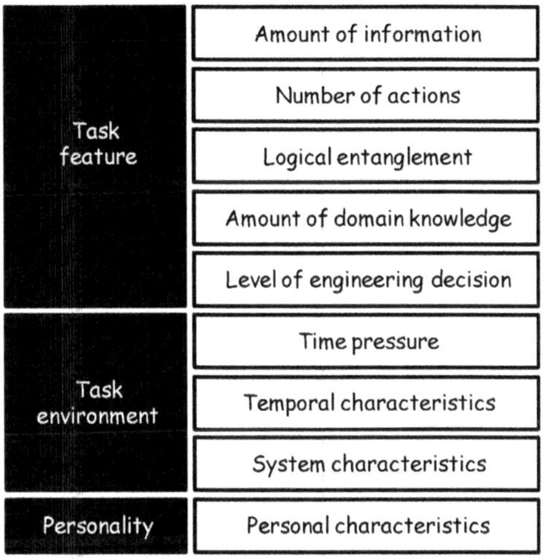

Fig. 3.7 Three groups of task complexity factors

The first group contains several categories pertaining to task features that can be characterized from a proceduralized task itself. For example, the amount of information as well as the number of actions can be easily obtained after a proceduralized task has been determined. In addition, it is expected that three categories of complexity factors, such as logical entanglement, the amount of domain knowledge, and the level of an engineering decision, can be extracted from the given proceduralized task. This strongly suggests that there will be a deterministic framework by which the effect of complexity factors on the performance of proceduralized tasks can be dealt with.

In contrast, the second group seems to defy easy measurement in a deterministic framework because of the dynamic features of a task environment. That is, it is very difficult to measure the effect of a task arrival rate, which belongs to the category of *temporal characteristics*, on the performance of proceduralized tasks because it would vary in the form of a continuous as well as a cumulative pattern over time. In addition, this effect would likely vary with respect to time constraints (e.g., time pressure). Accordingly, a stochastic framework would be necessary to reflect the varied effects of complexity factors belonging to these categories. Similarly, due to the diversity of personalities, a stochastic framework should be used to ponder the effect of a personality on the performance of proceduralized tasks.

Consequently, in measuring the effect of complexity factors on the performance of proceduralized tasks, it is reasonable to start from easy and tangible features first. Therefore, five categories of complexity factors that are closely related to *task features* are worth considering first. This implies that the systematic framework we are trying to develop can be regarded as a kind of static (as well as objective) complexity measure. That is, if it is possible to characterize all five complexity factors without reference to any dynamic (e.g., temporal characteristics) or subjective (e.g., personal characteristics) constituents, the result of the developed framework would represent the *verbatim complexity* of a proceduralized task to be given to every qualified operator who has to accomplish it.

References

Allen JA, Teague RC, Carter RE (1996) The effects of network size and fault intermittency on troubleshooting performance. IEEE Trans Syst Man Cybern A: Syst Hum 26(1):125–132

Boy GA, Brito G (2000) Toward a categorization of factors related to procedure following and situation awareness. In: International Conference on Human Computer Interaction in Aeronautics (HCI-Aero'00), Toulouse

Campbell DJ (1988) Task complexity: a review and analysis. Acad Manage Rev 13(1):40–52

Campbell DJ, Gingrich KF (1986) The interactive effects of task complexity and participation on task performance: a field experiment. Organizat Behav Hum Decis Processes 38:162–180

Ditlevsen O (2003) Decision modeling and acceptance criteria. Structu Saf 25:165–191

Jacko JA, Salvendy G (1996) Hierarchical menu design: breadth, depth and task complexity. Percept Mot Skills 82:1187–1201

Jiang JJ, Klein G (2000) Side effects of decision guidance in decision support systems. Interact Comput 12:469–481

Johannsen G, Levis AH, Stassen HG (1994) Theoretical problems in man-machine systems and their experimental validation. Automatica 30:217–231

Johnson EJ, Payne JW (1985) Effort and accuracy in choice. Manage Sci 31:395–414

Jonassen DH (2000) Toward a design theory of problem solving. Educat Technol Res Develop 48(4):63–85

Kieras D, Polson PG (1985) An approach to the formal analysis of user complexity. Int J Man-Machine Stud 22:365–394

Leitch R, Gallanti M (1992) Task classification for knowledge-based systems in industrial automation. IEEE Trans Syst Man Cybern 22:142–152

Leplat J (1998) Task complexity in work situations. In: Goodstein LP, Anderson HB, Olsen SE (eds) Tasks, Errors and Mental Models, Taylor and Francis, London, pp.105–115

Li K, Wieringa PA (2000) Understanding perceived complexity in human supervisory control. Cognit Technol Work 2:75–88

Liao C, Palvia PC (2000) The impact of data models and task complexity on end-user performance: an experimental investigation. Int J Hum-Comput Stud 52:831–845

Maynard DC, Hakel MD (1997) Effects of objective and subjective task complexity on performance. Hum Perform 10(4):303–330

Morris NM, Rouse WB (1985) Review and evaluation of empirical research in troubleshooting. Hum Factors 27(5):503–530

Park J, Jung W, Jung K (2008) The effect of two complexity factors on the performance of emer-

gency tasks - an experimental verification. Reliabil Eng Syst Saf 93:350–362
Rouse WB (1978) Human problem solving performance in a fault diagnosis task. IEEE Trans Syst Man Cybern 8(4):258–271
Rouse WB, Rouse SH (1979) Measures of complexity of fault diagnosis task. IEEE Trans Syst Man Cybern 9(11):720–727
Schmuck P Gundlach W (1989) Reduction of mental effort in tasks of different complexity. In: Klix F, Streitz NA, Waern Y, Wandke H (eds) Man-computer Interaction Research. Elsevier, Amsterdam
Shugan SM (1980) The cost of thinking. J Consumer Res 7(2):99–111
Sintchenko V, Coiera E (2002) Which clinical decision benefit from automation? A task complexity approach. In: Surjan G, Engelbrecht R, McNair P (eds) Proceedings of MIE2002, IOS Press, Amsterdam, pp.639–648
Spangler WE and Peters JM (2001) A model of distributed knowledge and action in complex systems. Decis Support Syst 31:103–125
Stassen HG, Johannsen G, Moray N (1990) Internal representation, internal model, human performance model and mental workload. Automatica 26(4):811–820
Sundstrom GA (1993) Towards models of tasks and task complexity in supervisory control applications. Ergonomics 36:1413–1423
Svensson E, Angelbrog-Thandrez M, Sjoberg L, Olsson S (1997) Information complexity – mental workload and performance in combat aircraft. Ergonomics 40:362–380
Thelwell PJ (1994) What defines complexity? In: Robertson SA (ed) Contemporary Ergonomics: Ergonomics for All, Taylor and Francis, London, pp.89–94
Turk Z (2001) Multimedia: providing students with real world experiences. Automat Construction 10(2):247–255
Van Eekhout JM, Rouse WB (1981) Human errors in detection, diagnosis and compensation for failures in the engine control room of a super tanker. IEEE Trans Syst Man Cybern 11(12):813–816
Wright P, Pocock S, Fields B (1998) The prescription and practice of work on the flight deck. In: Green TRG, Bannon L, Warren CP, Buckley J (eds) 9th European Conference on Cognitive Ergonomics, Limerick University Press, Limerick, Ireland, pp.37–42
Wood RE (1986) Task complexity: definition of the construct. Organizat Behav Hum Decis Processes 37:60–82
Wood RE, Locke EA (1990) Goal setting and strategy effects on complex tasks. Res Organizat Behav 12:73–109

Part II
Complexity Evaluation

4 Introduction to Software Complexity

At the end of the previous chapter, five categories of complexity factors that would serve as a starting point to deterministically evaluate the complexity of proceduralized tasks were identified. As In this chapter, software complexity measures will be explained as a theoretical basis for quantifying the complexity of proceduralized tasks. In this regard, it may be necessary to start this chapter by examining why software complexity must be considered in order to quantify the complexity of proceduralized tasks.

4.1 Software Complexity

We live in a very convenient time, and our lives are made easier by various kinds of computer technologies. For example, (1) we can buy a book from an online bookstore managed by powerful mainframes as well as sophisticated software, (2) we can produce merchandise using a fully automated machine controlled by well-structured software, and (3) we can even operate on a patient using a robot that is manipulated by precise software. However, in order to enjoy these conveniences we must secure reliable software that is able to perform all the required functions we want. For this reason, allied industries have been spending a tremendous amount of money and other resources to develop reliable software.

From this standpoint, one of the canonical approaches is to manage the complexity of software, because it directly affects software maintainability. Carver (1987) pointed out that the maintainability of software is a kind of quantitative measure that makes it possible to evaluate how easy it is to understand given software. Similarly, Gibson and Senn (1989) stated that *maintainability is defined as the ease with which systems can be understood and modified* (p. 348). Although there are other definitions about maintainability, it is evident that maintenance is one of the crucial aspects determining the reliability of software, because it contains all kinds of software engineering activities required after the implementation of software. Carver (1987) summarized these activities as follows:

> These distinct categories of maintenance can be identified: (1) corrective maintenance, (2) adaptive maintenance, and (3) perfective maintenance. Corrective maintenance is the diagnosis and correction of latent software errors. It is required when errors undiscovered during testing and debugging are found. Since a correct program is rare, latent errors are common. The errors may vary in impact from trivial to critical. In any case, the code must

be modified to correct the error. Adaptive maintenance is maintenance due to changes in the external environment of a program. New generations of hardware and later releases of software are among causes of adaptive maintenance. Perfective maintenance is maintenance intended to enhance the system to meet the changing needs of the user. It includes modifications of existing functions, inclusion of general enhancements, and modifications for improved system performance (p. 299).

Therefore, if a new error is introduced in the course of performing software maintenance activities, the increase in maintenance costs is unavoidable (Cant et al. 1995; Carver 1987; Gibson and Senn 1989; Hops and Sherif 1995; Lew et al. 1988; Soi 1985). A more serious problem is that the possibility of undesired consequences will increase in proportion to the increase of the possibility of software malfunctions. As a result, since the early 1970s, diverse research projects on software complexity have been conducted in order to quantitatively control as well as predict the complexity of software, because it has been revealed that maintenance personnel are apt to show impaired performance when they have to deal with complicated software (Curtis et al. 1979; Davis and LeBlanc 1988; Kafura and Reddy 1987; McCabe and Butler 1989; Rombach 1987).

It is worth emphasizing that one of the major purposes of quantifying the complexity of software is to evaluate its understandability. For example, Gibson and Senn (1989) stated that *the more complex system is, the more difficult it is to understand, and therefore to maintain* (p. 347). Similarly, Carver (1987) pointed out that "Ease of understanding decreases as program complexity increases. Since complexity is a measure of the effort to comprehend, to maintain and to test software, the level of complexity of a program affects the maintainability of a program (p. 299)."

Moreover, Davis and LeBlanc (1988) articulated that "Available evidence and the opinion of many experts strongly suggest that programmers do not understand programs on a character by character basis. Rather they assimilate groups of statements which have a common function (p. 1366)."

This means that a theoretical framework quantifying the complexity of software can be used for quantifying the complexity of proceduralized tasks because (1) software complexity mainly deals with the level of understandability of software and (2) understandability in software complexity focuses not on reading comprehension (i.e., WR, CR and CMP) but on task comprehension, which affects the performance of tasks to be done by qualified operators (i.e., TP). Actually, this is not a new idea, because other researchers have already tried to apply software complexity measures to evaluating the complexity of supervisory control tasks (Murray and Liu 1994) and vice versa (Darcy et al. 2005). Therefore, it is very helpful to scrutinize the applicability of software complexity measures to quantifying the complexity of proceduralized tasks.

4.2 Software Complexity Measure

Many kinds of unique measures that are capable of quantifying the complexity of software from diverse viewpoints have been suggested for several decades. However, without loss of generality, software complexity measures fall into one of the following four categories: (1) those based on the size of the software, (2) those based on the data structure of the software, (3) those based on the control structure of the software, and (4) a combination of the first three measures (Carver 1987; Coskun and Grabowski 2001; Davis and LeBlanc 1988; Fenton and Neil 1999; Gonzalez 1995; Hops and Sherif 1995; Huang and Lai 1998; Khoshgoftaar et al. 1997; Lakshmanan et al. 1991; Soi, 1985).

First, one of the representative measures belonging to the first category is the line of code (LOC). This measure is very clear and straightforward because it is strongly expected that the longer the software source code, the greater the complexity of the software. Another typical measure is Halstead's E measure, which considers the frequencies of occurrence of operators as well as operands included in source code. Figure 4.1 illustrates how to quantify the value of Halstead's E measure with respect to an arbitrary source code.

Source code	Operator	Frequency	Operand	Frequency
IF (A = 0) THEN	;	2	A	3
A = B;	=	3	B	1
ELSE	( )	1	C	1
A = C;	IF	1		
	THEN	1		
	ELSE	1		

$$E = \frac{\eta_1 N_2 (N_1 + N_2)}{2\eta_2} \log_2(\eta_1 + \eta_2) = 221.9$$

η_1 = Number of unique operators = 6

η_2 = Number of unique operands = 3

N_1 = Total number of operators = 9

N_2 = Total number of operands = 5

Fig. 4.1 Quantifying the value of Halstead's E measure (Park et al. 2001, © Elsevier)

Second, the complexity of software can be quantified from the point of view of a data structure. Regarding this, it would be interesting to quote Wirth (1985):

> Yet, it is abundantly clear that a systematic and scientific approach to program construction primarily has a bearing in the case of large, complex programs which involve complicated sets of data. Hence, a methodology of programming is also bound to include all aspects of data structuring. Programs, after all, are concrete formulations of abstract algorithms based on particular representations and structures of data (p. 7).

This strongly suggests that complicated software requires complicated data structures as well as huge amounts of data. Accordingly, the complexity of data structures should be a good measure for quantifying the complexity of software. For this reason, many kinds of complexity measures that are able to deal with the complexity of data structures have been suggested. One of the typical measures is the depth of a data structure graph (Gonzalez 1995). Here, data structure graph means a graph that consists of nodes and arcs, where nodes denote data entities and arcs represent the relationship between nodes. For example, the hierarchical level of an arbitrary data structure shown in Fig. 4.2 is three due to the existence of a linear array (refer to the area surrounded by dotted lines).

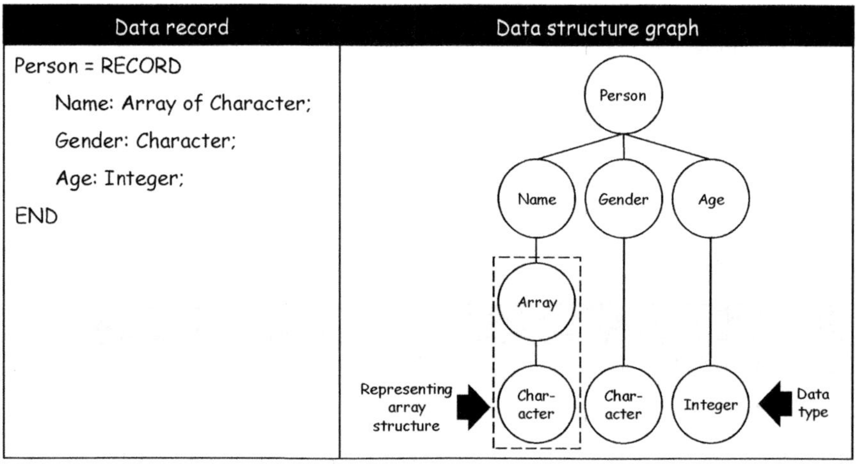

Fig. 4.2 Example of a data structure graph

Third, much work has been done considering the effect of a control flow graph on the complexity of software. Here, control flow graph (also called a program control graph) means a directed graph that has a unique entry and exit node, which is very similar to the flowchart of software (Baker 1978; Lakshmanan et al. 1991). In a control flow graph, each node denotes a block in source code that performs a specific function, and each arc represents a branch taken between nodes (Ramamurthy and Melton 1988). Therefore, it is very straightforward to expect that the complexity of software will be proportional to the complexity of the control flow graph.

One of the canonical measures belonging to this category is McCabe's cyclomatic complexity (v), which can be calculated by $v=e-n+2p$. Here, e, n, and p denote the number of edges (i.e., arcs), the number of nodes, and the number of connected components included in an arbitrary control flow graph, respectively. More simply, it was found that v is equal to the number of decision nodes plus one (McCabe and Butler; 1989). Therefore, from the point of view of McCabe's cyclomatic complexity, the complexity of two control flow graphs shown in Fig. 4.3 is identical, because they have two decision nodes.

Lastly, it is possible to measure the complexity of software by combining two or more complexity measures that belong to the aforementioned categories. For example, Ramamurthy and Melton (1988) and Curtis et al. (1979) suggested novel measures based on the integration of Halstead's E measure with McCabe's cyclomatic complexity. In addition, Bail and Zelkowitz (1988) and Oviedo (1980) suggested software complexity measures by simultaneously considering the control flow graph and the data structure graph of software.

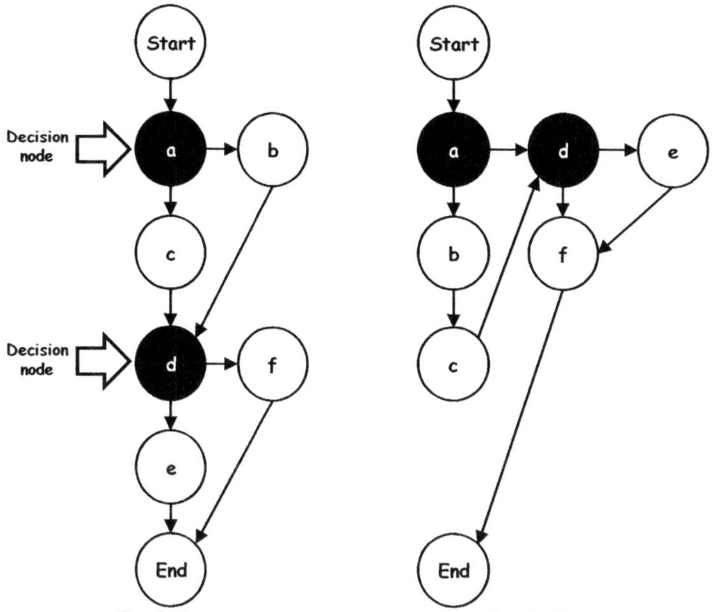

Graph G Graph G′

Fig. 4.3 Two control flow graphs with the same McCabe's cyclomatic complexity

Here, it is important to point out that there is another complexity measure that belongs to this category. That is, instead of combining several complexity measures that quantify the complexity of software using *different* methods, a new measure can be developed based on the integration of submeasures quantifying the complexity of software with an *identical* method. A typical example is a measure based on the concept of graph entropies because, as illustrated in Figs. 4.2 and 4.3, many graphic representation techniques have been used to analyze the characteristics of software.

4.3 The Concept of Graph Entropies

Traditionally, the entropy concept has been widely adopted in various research areas because it is very useful for expressing the degree of complexity (Shannon 1948). For this reason, including a series of works done by Mowshowitz (Mowshowitz 1968a-d), many researchers have expended considerable effort to quantify the complexity of software using the concept of graph entropies (Davis and LeBlanc 1988; Huang and Lai 1998; Gonzalez 1995; Lew et al. 1988). For example, let us consider the definition of the first-order and the second-order entropy suggested by Davis and LeBlanc (1988).

In order to quantify the first order entropy, the classes of nodes in a control flow graph should be identified based on their in- and out-degree as they appear. If there are nodes that share the same in- and out-degree, then they are regarded as nodes belonging to an equivalent class. In this regard, Fig. 4.4 depicts how to quantify the first-order entropy of two arbitrary graphs shown in Fig. 4.3.

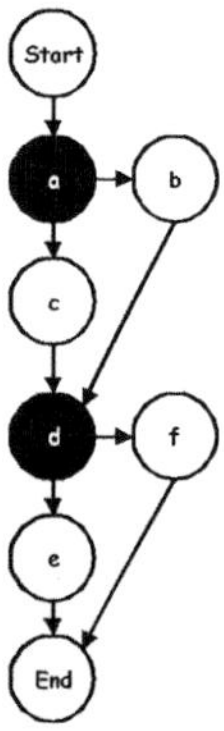

Class	In-degree	Out-degree	Node
I	0	1	{Start}
II	1	1	{b, c, e, f}
III	1	2	{a }
IV	2	0	{End}
V	2	2	{d}

A_i =Number of nodes belonging to the ith distinctive class

N =Total number of nodes in a graph

h =Number of distinctive classes

p_i = Estimated probability of the ith distinctive class = A_i/N

The first-order entropy of graph G = $H_1(G) = -\sum_{i=1}^{h} p_i \log_2 p_i$

$$= -\left\{\frac{1}{8}\log_2(\frac{1}{8}) + \frac{4}{8}\log_2(\frac{4}{8}) + \frac{1}{8}\log_2(\frac{1}{8}) + \frac{1}{8}\log_2(\frac{1}{8}) + \frac{1}{8}\log_2(\frac{1}{8})\right\} = 2.000$$

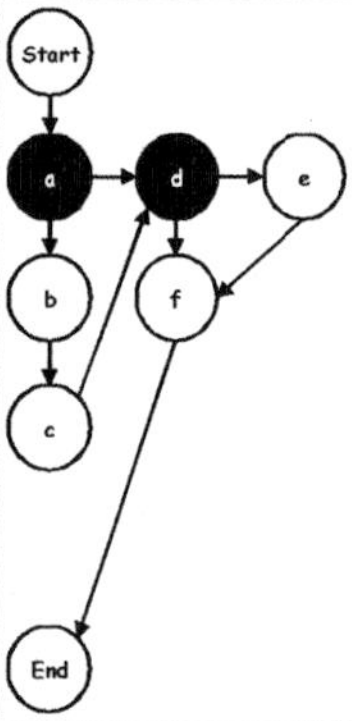

Class	In-degree	Out-degree	Node
I	0	1	{Start}
II	1	0	{End}
III	1	1	{b, c, e}
IV	1	2	{a}
V	2	1	{f}
VI	2	2	{d}

The first-order entropy of graph G′ = $H_1(G')$

$$= -\left\{\frac{1}{8}\log_2(\frac{1}{8}) + \frac{1}{8}\log_2(\frac{1}{8}) + \frac{3}{8}\log_2(\frac{3}{8}) + \frac{1}{8}\log_2(\frac{1}{8}) + \frac{1}{8}\log_2(\frac{1}{8}) + \frac{1}{8}\log_2(\frac{1}{8})\right\}$$

$$= 2.406$$

Fig. 4.4 The first order entropy of two arbitrary control flow graphs

In Fig. 4.4, it is apparent that all the nodes included in a graph G fall into the following classes: {Start}, {b, c, e, f}, {a}, {End}, and {d}. Accordingly, the number of distinctive classes denoted by h is five. In addition, the probability of each class is 1/8, 4/8, 1/8, 1/8, and 1/8, respectively. In this way, h and the probability of the associated classes can be calculated with respect to the graph G′. As a result, the first-order entropy of graphs G and G′ is 2.000 and 2.406, respectively.

From the point of view of the logical entanglement of control flow graphs, the value of the first-order entropy is very interesting, because the logic structure of graph G′ seems to be more complicated than that of graph G. Intuitively, this result is meaningful, because a control flow graph that consists of many equivalent nodes will tend to have a lower first-order-entropy value. In other words, if there is a kind of regularity in a control flow graph, it is expected that the value of the first-order entropy will be reduced because of the repetition of similar execution patterns, which results in an increase of the number of nodes belonging to identical node classes. In contrast, the value of the first-order entropy will increase due to irregular execution patterns, because the number of distinctive classes that are necessary to express the irregularity of execution patterns will increase. This means that the effect of logical entanglement on the complexity of software can be quantified by the first-order entropy.

Similarly, the second-order entropy can be calculated except for the class identification scheme. That is, nodes are considered to be equivalent if they share identical neighbors within one arc distance. The intention of this classification scheme is to express the amount of information that is needed to describe each node position, since the comprehension of a control flow graph becomes difficult with respect to the increase in the number of distinctive classes. For example, let us consider Table 4.1, which shows the distinctive classes of two control flow graphs G and G′, which are necessary to calculate the values of the second-order entropy.

Table 4.1 Distinctive classes of two control flow graphs

Graph G		Class	Graph G′	
Node	Neighbor node		Node	Neighbor node
{Start}	{a}	I	{Start}	{a}
{a}	{Start, b, c}	II	{a}	{Start, b, d}
{b, c}	{a, d}	III	{b}	{a, c}
{d}	{b, c, e, f}	IV	{c}	{b, d}
{e, f}	{d, End}	V	{d}	{a, c, e, f}
{End}	{e, f}	VI	{e}	{d, f}
–	–	VII	{f}	{d, e, End}
–	–	VIII	{End}	{f}

Based on the results of node class identifications summarized in Table 4.1, the values of the second-order entropy of two graphs can be calculated as below.

The second-order entropy of graph G = $H_2(G)$

$$= -\sum_{i=1}^{6} p_i \log_2 p_i$$

$$= -\left\{\frac{1}{8}\log_2(\frac{1}{8}) + \frac{1}{8}\log_2(\frac{1}{8}) + \frac{2}{8}\log_2(\frac{2}{8}) + \frac{1}{8}\log_2(\frac{1}{8}) + \frac{2}{8}\log_2(\frac{2}{8}) + \frac{1}{8}\log_2(\frac{1}{8})\right\} = 2.500$$

The second-order entropy of graph G′ = $H_2(G')$

$$= -\sum_{i=1}^{8} p_i \log_2 p_i$$

$$= -\left\{\begin{matrix}\frac{1}{8}\log_2(\frac{1}{8}) + \frac{1}{8}\log_2(\frac{1}{8}) + \frac{1}{8}\log_2(\frac{1}{8}) + \frac{1}{8}\log_2(\frac{1}{8}) \\ + \frac{1}{8}\log_2(\frac{1}{8}) + \frac{1}{8}\log_2(\frac{1}{8}) + \frac{1}{8}\log_2(\frac{1}{8}) + \frac{1}{8}\log_2(\frac{1}{8})\end{matrix}\right\} = 3.000$$

As can be seen from the above results, the value of the second-order entropy will increase in proportion to the increase in the number of nodes because the meaning of each node position becomes more unique. This means that more effort is required to understand the contents of software that consists of many nodes. Therefore, the second-order entropy of a control flow graph can be used to measure the effect of size on the complexity of software.

Here, it should be noted that the second-order entropy can be used to quantity the amount of information pertaining to a data structure graph. In other words, if the second-order entropy of an arbitrary graph implies the amount of information needed to understand its contents, the second-order entropy of a data structure graph can be used to measure the effect of the data structure on the complexity of the software. Therefore, based on the concept of graph entropies, it is possible to define a novel measure of software complexity. For example, Lew et al. (1988), Gonzalez (1995), and Huang and Lai (1998) proposed a novel measure by integrating several complexity measures quantified by the concept of graph entropies.

4.4 Selecting Appropriate Measures

At the end of Sect. 4.1, it was pointed out that software complexity measures could be used for quantifying the complexity of proceduralized tasks. The easiest way to do this is to use the associated software complexity measure that is able to evaluate one of the task complexity factors. For example, let us look at Table 4.2, which compares five kinds of task complexity factors with the associated software complexity measures.

Table 4.2 Comparing task complexity factors with the associated software complexity measures

Task complexity factor	Software complexity measure based on	Example
Amount of information	Data structure of software	The hierarchical level of a data structure graph
Number of actions	Size of software	Halstead's E measure
Logical entanglement	Control structure of software	McCabe's cyclomatic complexity
Amount of domain knowledge	–	–
Level of engineering decision	–	–

Table 4.2 suggests that we should be able to use Halstead's E measure to quantify the effect of the number of actions on the complexity of proceduralized tasks, because this related to the size of software, which would be directly comparable to the size of proceduralized tasks (i.e., number of actions to be conducted by qualified operators). Similarly, McCabe's cyclomatic complexity, which evaluates the logical entanglement of the control structure of software, would be a good alterna-

tive to quantify the effect of logical entanglement on the complexity of proceduralized tasks.

Unfortunately, there are three critical problems in this approach. First, there is no corresponding software complexity measure that is capable of evaluating the effect of the amount of domain knowledge on the complexity of proceduralized tasks. Likewise, there is no appropriate software complexity measure regarding the level of engineering decision.

Second, even if corresponding software complexity measures were available, some of them would likely have limited application to the quantification of the complexity of proceduralized tasks. For example, let us recall the value of the first-order entropy of two arbitrary graphs G and G′ (Fig. 4.3). From the point of view of McCabe's cyclomatic complexity, these two graphs have the same value. However, it is intuitively evident that the control structure of graph G′ is more complicated than that of graph G. This means that there are times when McCabe's cyclomatic complexity is not appropriate for quantifying the effect of logical entanglement on the complexity of proceduralized tasks. In addition, the result of a previous study has revealed that Halstead's E measure has a limitation in application to the complexity of proceduralized tasks (Park et al. 2001).

This limitation engenders the third problem, which is related to integrating the effects of five kinds of task complexity factors. It is very natural to assume that the overall complexity of proceduralized tasks should be determined based on the integration of partial contributions originating from five kinds of task complexity factors. Unfortunately, this is not a valid idea. Let us assume that we quantified the effects of the number of actions and logical entanglement on the complexity of proceduralized tasks by using Halstead's E measure and McCabe'c cyclomatic complexity, respectively. Nevertheless, combining the value of Halstead's E measure with that of McCabe's cyclomatic complexity is less meaningful because, as mentioned earlier, there are times when these measures give inappropriate results about the complexity of proceduralized tasks. In addition, the integration of heterogeneous measures would become another source of difficulty in quantifying the complexity of proceduralized tasks.

For the above reasons, a better way to quantify the complexity of proceduralized tasks seems to use the concept of graph entropies. That is, if we construct a series of graphs that are able to represent the nature of five kinds of task complexity factors, the contribution of each factor can be quantified by either the first-order entropy or the second-order entropy. In addition, since the technical basis of graph entropies is homogeneous to some extent (i.e., the entropy value of an arbitrary graph can be calculated by a set of probabilities obtained from the definition of a node classification scheme), it is expected that one should be able to integrate the contributions of five kinds of task complexity factors into a single and meaningful value.

References

Bail WG, Zelkowitz MV (1998) Program complexity using hierarchical computers. Comput Lang 13(3/4):109–123
Baker TP (1978) Natural properties of flowchart step-counting measures. J Comput Syst Sci 16:1–22
Cant SN, Jeffery DR, Henderson-Sellers B (1995) A conceptual model of cognitive complexity of elements of the programming process. Inf Softw Technol 37(7):351–362
Carver DL (1987) Producing maintainable software. Comput Ind Eng 12(4):299–305
Coskun E, Grabowski M (2001) An interdisciplinary model of complexity in embedded intelligent real-time systems. Inf Softw Technol 43:527–537
Curtis B, Sheppard SB, Milliman P, Borst MA, Love T (1979) Measuring the psychological complexity of software maintenance tasks with the Halstead and McCabe Metrics. IEEE Trans Softw Eng 5(2):96–104
Darcy DP, Kemerer CF, Slaughter SA, Tomayko JE (2005) The structural complexity of software: An experimental test. IEEE Trans Softw Eng 31(11):982–995
Davis JS, LeBlanc RJ (1988) A study of the applicability of complexity measures. IEEE Trans Softw Eng 14(9):1366–1372
Fenton NE, Neil M (1999) Software metrics: successes, failures and new directions. J Syst Softw 47:149–157
Gibson VR, Senn JA (1989) System structure and software maintenance performance. Commun ACM 32(3):347–358
Gonzalez RR (1995) A unified metric of software complexity: measuring productivity, quality and value. J Syst Softw 29:17–37
Hops JM, Sherif JS (1995) Development and application of composite complexity models and a relative complexity metric in a software maintenance environment. J Syst Softw 31:157–169
Huang SJ, Lai R (1998) On measuring the complexity of an Estelle specification. J Syst Softw 40:165–181
Kafura D, Reddy GR (1987) The use of software complexity metrics in software maintenance. IEEE Trans Softw Eng 13(3):335–343
Khoshgoftaar TM, Allen EB, Lanning DL (1997) An information theory-based approach to quantifying the contribution of a software metric. J Syst Softw 36:103–113
Lakshmanan KB, Jayaprakash S, Sinha PK (1991) Properties of control-flow complexity measures. IEEE Trans Softw Eng 17(12):1289–1295
Lew KS, Dillon TS, Forward KE (1988) Software complexity and its impact on software reliability. IEEE Trans Softw Eng 14(11):1645–1655
McCabe TJ, Butler CW (1989) Design complexity measurement and testing. Commun ACM 32(12):1415–1425
Mowshowitz A (1968a) Entropy and the complexity of graphs: I. An index of the relative complexity of a graph. Bull Math Biophys 30:175–204
Mowshowitz A(1968b) Entropy and the complexity of graphs: II. The information content of digraphs and infinite graphs. Bull Math Biophys 30:225–240
Mowshowitz A (1968c) Entropy and the complexity of graphs: III. Graphs with prescribed information content. Bull Math Biophys 30:387–414
Mowshowitz A (1968d) Entropy and the complexity of graphs: IV. Entropy measures and graphical structure. Bull Math Biophys 30:533–546
Murray J, Liu Y (1994) A software engineering approach to assessing complexity in network supervision tasks. In: Proceedings of the IEEE International Conference on Human, Information and Technology, San Antonio, TX, 1:25–29

Oviedo EI (1980) Control flow, data flow and program complexity. In: Proceedings on IEEE COMPSAC, Chicago, pp.146–152,

Park J, Jung W, Ha J (2001) Development of the step complexity measure for emergency operating procedures using entropy concepts. Reliabil Eng Syst Saf 71:115–130

Ramamurthy B, Melton A (1988) A synthesis of software science measures and the cyclomatic number. IEEE Trans Softw Eng 14(8):1116–1121

Rombach HD (1987) A controlled experiment on the impact of software structure on maintainability. IEEE Trans Softw Eng 13(3):344–354

Shannon CE (1948) A mathematical theory of communication. Bell Syst Tech J 27:379-423/623–656

Soi IM (1985) Software complexity: an aid to software maintainability. Microelectron Reliabil 25(2):223–228

Wirth N (1985) Algorithms and Data Structures. Prentice Hall, Englewood Cliffs, NJ

5 Emergency Tasks Prescribed in the EOPs of NPPs

In Chap. 3, we identified five kinds of complexity factors that can complicate the performance of proceduralized tasks. In addition, a theoretical basis to quantify the complexity of proceduralized tasks was explained in Chap. 4. Therefore, the next phase is to develop a quantification method that is able to calculate the contribution of each complexity factor. First, it would be worthwhile to review the characteristics of emergency tasks prescribed in the EOPs of NPPs, because detailed explanations about the quantification method will be described based on them.

5.1 Design Features of Pressurized Water Reactors

According to the recent statistics of the Nuclear Energy Institute (NEI), a total of 448 NPPs are under commercial operation in 30 countries as of April 2008 (NEI 2008). In addition, nine different types of NPPs are now operating all over the world. They are (1) advanced boiling light-water-cooled and moderated reactor (ABWR), (2) advanced gas-cooled, graphite-moderated reactor (AGR), (3) boiling light-water-cooled and moderated reactor (BWR), (4) fast breeder reactor (FBR), (5) gas-cooled, graphite-moderated reactor (GCR), (6) light-water-cooled, graphite-moderated reactor (LWR), (7) pressurized heavy-water-moderated and cooled reactor (PHWR), (8) pressurized light-water-moderated and cooled reactor (PWR), and (9) water-cooled-water-moderated power reactor (WWER). Figure 5.1 shows the simplified schematic of a PWR that is the most popular NPP in the world. For more information about the design as well as supporting systems of PWRs, please refer to fundamental information provided by NSIC (2008) or USNRC (2008). Well-known Web sites such as AKIP (2008) or Virtual Nuclear Tourist (2008) are also good sources of basic information about commercial NPPs.

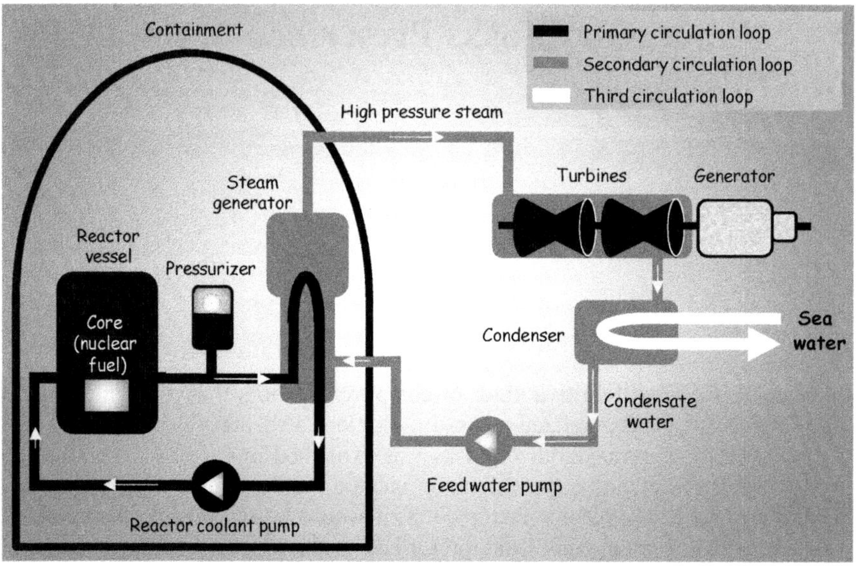

Fig. 5.1 Simplified schematic of a PWR

The backbone of PWRs is the primary circulation loop, usually called a reactor coolant system (RCS), which generally contains a pressurizer, reactor coolant pumps (RCPs), and steam generators (SGs). The RCS connects to a reactor vessel so that thermal energy generated from the nuclear fission of a core (nuclear fuel assemblies) heats up water in the RCS (i.e., the primary coolant) from a temperature of about 300°C to 320°C (572°F and 608°F, respectively). To this end, the pressurizer maintains the pressure of the RCS within from 1.2×10^7Pa to 1.6×10^7Pa (1740 psi and 2320 psi, respectively) in order to prevent the boiling of the primary coolant (which is why this NPP is called a PWR). Then the heated primary coolant is pumped to the SGs in order to generate steam by transferring the heat of the RCS to the coolant of the secondary circulation loop (i.e., the secondary coolant). To facilitate this process, each SG contains many inverted U-shape tubes (from about 3000 to 16000) having a very small diameter (about 19 mm or 3/4 in). That is, the primary coolant passing through the inside the tubes transfers the heat to the secondary coolant passing outside the tubes. As a result, the secondary coolant becomes a high pressure steam in the SGs. This steam rotates the blades of turbines to generate electricity from a generator. Then, in condensers that are very large heat exchangers cooled by sea water, river water, or air, the exhausted steam is condensed into water. Finally, in order to reheat the water, a feed water pump transfers the condensed water to the SGs.

It should be emphasized that one of the unique design features of PWRs is three independent (or separated) circulation loops. For example, the RCS forms the primary circulation loop, while the turbines, condensers, and feed water pumps comprise the secondary circulation loop. In addition, the third circulation loop is necessary to condense the exhausted steam using an external heat sink, such as sea water. This means that radioactive materials produced as a result of a nuclear fission should be confined to the primary circulation loop. In addition, even if there is a breach in the primary circulation loop, the containment can effectively blockade the leakage of radioactive materials to the environment.

5.2 Event- and Symptom-based Procedures

As can be perceived from the name, EOPs consist of many procedures containing a set of proceduralized tasks to be done when an emergency event has occurred. In other words, emergency tasks prescribed in EOPs allow qualified operators to lead the condition of NPPs to an established operating boundary by providing practical actions to cope with an emergency event. In light of this concern, two types of EOPs have been used in PWRs for several decades. To understand the characteristics of EOPs, let us consider Fig. 5.2, which shows a hypothetical troubleshooting table including typical symptoms and associated diseases, which were collected from the Internet.

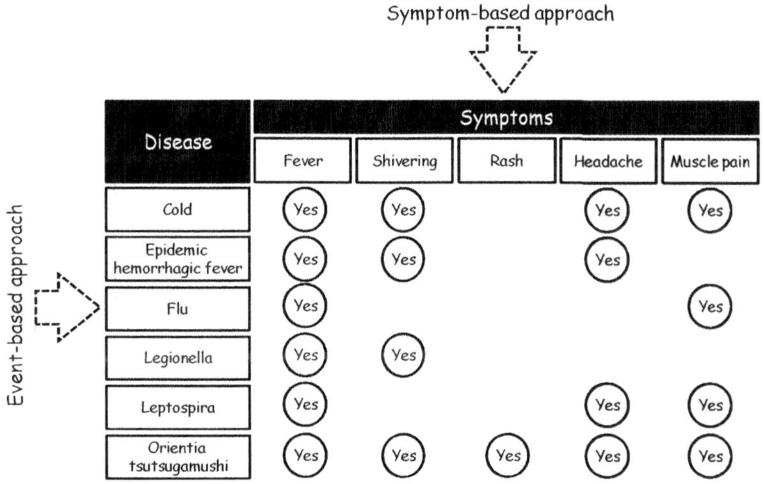

Fig. 5.2 Hypothetical troubleshooting table

Regarding this troubleshooting table, let us assume that we are trying to develop a series of medical procedures through which a less experienced physician (e.g., a qualified operator) determines what should be done to cope with the diseases of patients. To this end, we can imagine two kinds of unique approaches. The first one is to develop event-based (or event-oriented) procedures, in which detailed actions to be taken by the physician are precisely described. Actually, since this approach is very straightforward, it is expected that the physician may easily perform key actions to heal the patients, such as selecting proper medication, adjusting dosage, and determining an appropriate medication term, etc.

Unfortunately, there are at least three obstacles in applying this approach. First, the number of procedures will be proportional to the number of existing diseases. In other words, one cannot avoid developing extensive event-based procedures to support the physician.

The second obstacle is the accuracy of a medical diagnosis. That is, event-based procedures are meaningful only if the nature of diseases is correctly identified. In fact, however, less experienced physicians tend to make mistakes in their diagnosis.

The third obstacle is more serious for the patients, because there are times when a physician is not able to make a proper diagnosis. For example, it may be very difficult for the physician to identify the nature of diseases that have occurred simultaneously, or to identify the outbreak of a new disease in a short period of time. This implies that patients could be in a big trouble, because not only the appropriate medical treatment is likely to be delayed for a long time but also the physician is apt to prescribe wrong medical treatments. Therefore, we need to change our viewpoint to overcome these obstacles. Alternatively, it is possible to adopt a symptom-based (or symptom-oriented) approach.

The fundamental concept of a symptom-based approach is quite simple. Instead of an event-based procedure that directly deals with each disease, we develop a set of procedures that cover generic medical treatments for each symptom. For example, if patients have a fever, then a physician could follow a procedure that would include many kinds of detailed actions to alleviate it. Therefore, this approach has a definite advantage because the physician does not need to accurately identify the nature of diseases. Nevertheless, because of the following drawbacks, we must keep in mind the potential for abuse in the symptom-based approach.

The first drawback is that symptom-based procedures are inefficient as compared with event-based procedures when a physician did make the correct decision about a particular disease. This is also unavoidable, because the underlying strategy of symptom-based procedures is not to eliminate the cause of a disease but to maintain the vital condition of patients within an allowable boundary by alleviating critical symptoms.

The second drawback of a symptom-based approach is that this strategy impels a physician to prioritize observed symptoms. That is, when the physician observes two or more symptoms, he or she perhaps feels a frustration, wondering which is the most urgent symptom. This means that, without clear prioritization criteria, the physician is likely to feel a difficulty in carrying out symptom-based procedures.

Therefore, form a practical point of view, it would be a good idea if we combined the aforementioned approaches. In other words, event-based procedures can be used when the nature of emergency events is properly identified, while symptom-based procedures can be used when emergency events that are difficult to diagnose (such as multiple events or unknown events, etc.) have occurred. Actually, this idea is known as *a symptom-oriented and event-specific approach*, and it has been regarded as a radical concept for developing the EOPs of PWRs (IAEA 1985, 1998). Table 5.1 briefly compares the pros and cons of the event- and the symptom-based approaches (Park et al. 1995). It is to be noted that more detailed explanations about the symptom-oriented and event-specific approach will be given in the following sections.

Table 5.1 Comparing pros and cons about two different approaches

Approach	Advantage	Disadvantage
Event-based	• Easy to use • Provides detailed and straightforward recovery actions	• Too many procedures due to the subdivision of events • Requires correct diagnosis • No guideline about unknown or multiple (concurrent) events
Symptom-based	• Deals with unknown or multiple events by providing generic recovery actions that are independent of the cause of events • Allows a unified procedure that is applicable to many events	• Less effective when a single or an apparent event has occurred • Requires intensive education as well as training to change an operating philosophy

5.3 The Generic Structure of EOPs

The United States Nuclear Regulatory Commission (USNRC 1982) has defined EOPs as follows:

> EOPs are plant procedures that direct operators' actions necessary to mitigate the consequences of transients and accidents that have caused plant parameters to exceed reactor protection system set points or engineered safety feature set point, or other established limits (p. 3).

With regard to this definition, the International Atomic Energy Agency (IAEA) has suggested a set of functional requirements about EOPs (see p. 58 of IAEA 1998). Some of them are given below.

- The objective of EOPs is to return NPPs to a condition covered by normal procedures or a safe and stable shutdown condition.
- Expected emergency conditions should be identified and EOPs for dealing them should be prepared for use when required.
- Since emergencies may not follow anticipated patterns, EOPs should provide for sufficient flexibility of actions to accommodate variations, including multiple and sequential failures.

In order to fulfill these requirements, many countries have applied the symptom-oriented and event-specific approach to the development of EOPs, since, without loss of generality, emergency events fall into two categories (CEOG 1996; WOG 1987). The first category corresponds to emergency events that can be properly identified in an analytical way, including (1) interpreting theoretical models, (2) simulating thermohydraulic codes, and (3) investigating historical data. A typical emergency event belonging to this category is design basis accidents (DBAs). Here, we would expect that the unwanted consequence of DBAs (e.g., the release of radioactive materials into the environment) would be minimized by implementing an optimal set of event-based recovery actions if we could correctly identify the nature of the accidents. In other words, it is possible to prescribe effective recovery actions that successfully lead the status of PWRs to a stable as well as safe condition when we know the cause of an emergency event. Based on this premise, therefore, event-based procedures have been used for several decades to cover diagnosable events.

In contrast, in the case of an emergency event that belongs to the second category, it may be less meaningful to use event-based procedures, because the nature of such an emergency event is so complicated that we are not able to specify which event-based procedure is applicable. Typical examples belonging to this category are (1) multiple events that have concurrently or simultaneously occurred and (2) instrumentation failures that are likely to distort or even hide the nature of an emergency event. Accordingly, in order to cope with these kinds of emergency events, symptom-based procedures are necessary (Meyer et al. 1987). However, a practical problem still remains in developing symptom-based procedures, that is, a theoretical basis for identifying a set of symptoms to be monitored as well as for determining their priority. Consequently, the concept of critical safety functions (CSFs) was introduced in the early 1980s (Corcoran et al. 1981; Surman et al. 1984).

To sum up, CSFs define a list of crucial functions with their relative priorities, which are useful for preventing intolerable consequences due to emergency events. In addition, each CSF is linked to the associated process parameters by which its integrity can be determined. For example, although there are several distinctive lists of CSFs, Fig. 5.3 shows some typical CSFs as well as the associated process parameters (Corcoran et al. 1984; Kadak 1984; Wilkinson 1984). Therefore, a symptom-based procedure, whose purpose is to secure the integrity of a specific CSF, can be developed based on the associated process parameters (i.e., symptoms). For this reason, symptom-based procedures are frequently referred to as *function-based procedures* or *symptom-based function-related procedures* (Surman et al. 1984; Wilkinson 1984).

From the above explanations it is possible to simplify the generic structure of EOPs as depicted in Fig. 5.4. It is to be noted that several DBAs should be covered by symptom-based procedures, because they directly jeopardize the integrity of CSFs. In other words, since the loss of any CSF means the breach of a defense block that is essential in securing the safety of PWRs, the restoration of CSFs has a priority compared to the response of a DBA. For example, in the case of anticipated transient without scram (ATWS), a symptom-based procedure should be developed because such an event promptly jeopardizes the most important CSF in Fig. 5.3 – *reactivity control*.

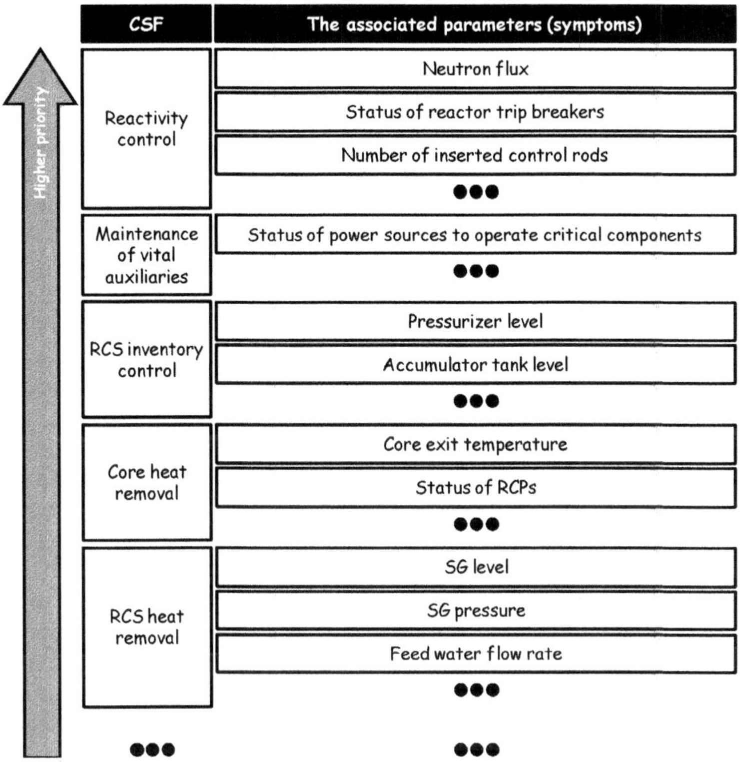

Fig. 5.3 Part of a typical CSF

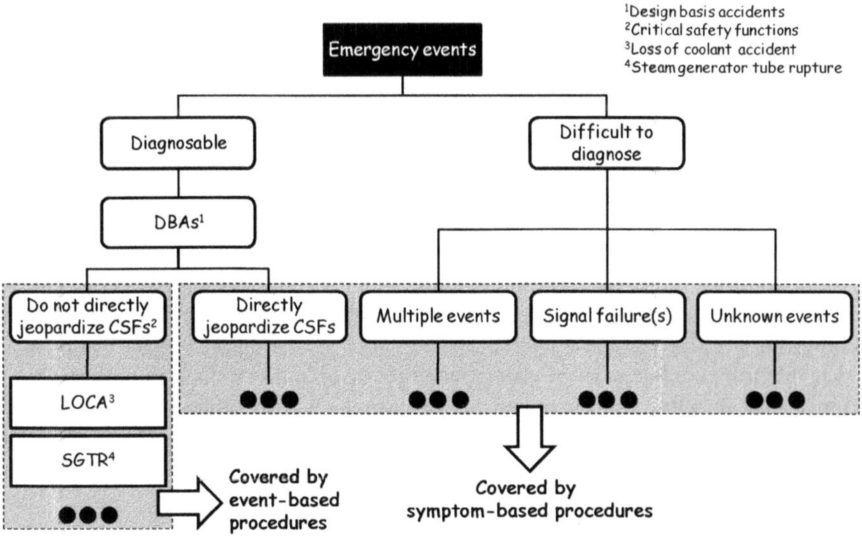

Fig. 5.4 The generic structure of EOPs (Park and Jung 2004, © Elsevier)

5.4 Emergency Tasks Prescribed in EOPs

Basically, either in event-based procedures or in symptom-based procedures, emergency tasks consist of one or more procedural steps including many actions to be conducted by qualified operators. For example, let us consider a steam generator tube rupture (SGTR) event that is a typical DBA for all kinds of PWRs including Korean standard nuclear power plants (KSNPs) (KHNP 2002).

The occurrence of SGTR denotes the breach of inverted-U tubes located in SGs. These tubes are very important because they constitute physical barriers between radioactive coolants circulating in the primary loop (i.e., the RCS) and non-radioactive coolants circulating in the secondary loop. This means that the integrity of the tubes is essential to minimize the leakage of radioactive coolants from the primary loop to the secondary loop. Otherwise, there is the potential that radioactive materials in the secondary loop could escape directly to the atmosphere in the form of steam. Therefore, it is necessary to systematically prepare emergency tasks so that the consequences of SGTR can be controlled at an acceptable level of risk.

To this end, let us think of several decisive symptoms that would appear when SGTR has occurred: (1) decreasing the amount of RCS coolants, (2) decreasing RCS pressure (3) increasing the water level of a ruptured SG (a SG with one or more ruptured tubes), (4) increasing a radioactivity level in the secondary circulation loop, etc. (CEOG 1996; WOG 1987). From these symptoms, emergency tasks to be performed by qualified operators can be determined on the basis of two criteria: (1) *which symptom should be urgently restored to an acceptable limit?* and (2) *how we can restore it*?

In this regard, it is possible to develop an optimal set of emergency tasks for SGTR. For example, we must give priority to the symptom of *decreasing the amount of RCS coolants* because it is directly related to *RCS inventory control*, which corresponds to the third CSF in Fig. 5.3. Consequently, when SGTR has occurred, the highest emergency task (except for the confirmation of occurrence of the SGTR) is to secure RCS inventory. For this reason, as depicted in Fig. 5.5, which shows some of the emergency tasks prescribed in the SGTR procedure of KSNPs, the fourth and fifth procedural steps constitute the second emergency task specifying how to secure the RCS inventory (Park et al. 2005).

In this way, if we can identify the cause of an emergency event, it is possible to operate the associated components or equipment in an optimal manner so as to restore the PWRs to a stable and safe state. This means that, to some extent, it is possible to (1) define a set of crucial emergency tasks to be done in a certain time limit (i.e., an allowable time) and (2) prepare a set of contingency actions to be carried out when preplanned instructions are not working. Actually, in order to achieve the second task, procedural steps are frequently presented in a two-column format.

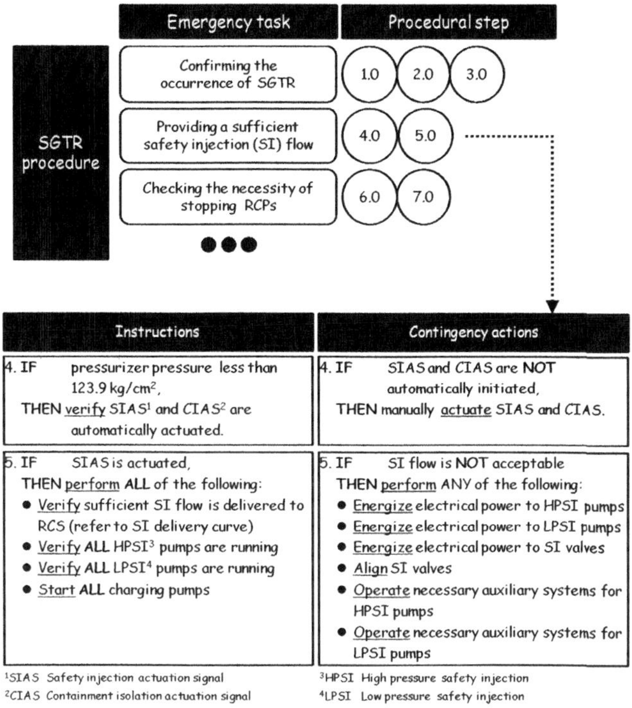

Fig. 5.5 Some emergency tasks prescribed in the SGTR procedure of KSNPs

The left column of Fig. 5.5 shows instructions that provide expected process responses, and the right column contains contingency actions that should be carried out if the instructions in the left column are not met. Accordingly, qualified operators are expected to move down and carry out actions in the left column if the expected responses are observed. In contrast, if the expected responses are not satisfied, qualified operators have to move to the right column in order to perform a set of contingency actions. After the contingency actions in the right column are successfully performed, qualified operators are expected to proceed to the remaining actions in the left column. This implies that qualified operators need to strictly follow the predefined sequence of actions. In this regard, Macwan and Mosleh (1994) classified four basic types of action sequences included in procedural steps, as illustrated in Fig. 5.6.

However, we need to at least consider one more action sequence that is related to the selection of equally acceptable actions. In order to understand the meaning of an equally acceptable action, let us consider the definition of equally acceptable steps (USNRC 1982).

> Equally acceptable steps are those for which any one of several alternative steps or sequence of steps may be equally correct. For these steps, the operator should always be directed to carry out one of the alternative steps (or sequences), but should also be given the other alternatives when it is possible that the designated steps (or sequence) cannot be done (e.g., a designated piece of equipment is unavailable) (p. 23).

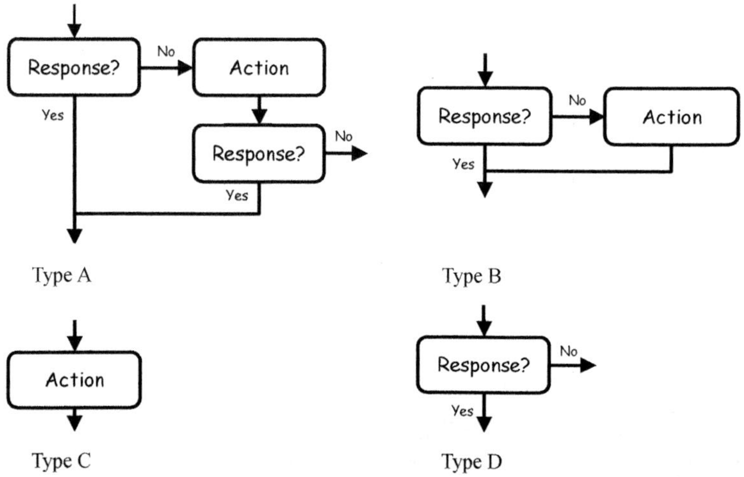

Fig. 5.6 The sequence of actions – four basic types (Macwan and Mosleh 1994, © Elsevier)

Along with this definition, we can define equally acceptable actions as *those for which any one of several alternative actions or sequence of actions may be equally correct.* Figure 3.4 shows a clear example of equally acceptable actions, because qualified operators have to select one of two plausible action sequences, either *increasing outflow* or *providing a bypass line*. In a situation in which qualified operators have to accomplish the required tasks by a certain time limit, equally acceptable actions could be a burden to them. In addition, even if there is sufficient time, it would not be easy to specify one action sequence because the evaluation of the pros and cons of all the plausible action sequences is mostly case sensitive. Similarly, the fifth procedural step in Fig. 5.5 contains equally acceptable actions. Nevertheless, to some extent, the use of equally acceptable actions seems to be unavoidable in the course of describing proceduralized tasks.

Let us recall the selection problem with the situations depicted in Fig. 3.6. As mentioned before, it is expected that most qualified operators should select the action sequence related to *providing a bypass line* when the water level is increasing drastically and CV 1 is 90% open. In this case, these equally acceptable actions can be reorganized as *Type B* of Fig. 5.6, such as *IF the water level of Tank 1 is increasing drastically AND CV 1 is 90% open, THEN provide the bypass line*. However, this action covers only a small part of all the situations with which qualified operators can be faced. In other words, there are no actions that are applicable to other situations, such as *increasing water level drastically and CV 1 is 10% open* or *gradually increasing water level CV 1 is 90% open,* etc. This means that it is very difficult (or even impossible) to specify detailed actions about each and every situation.

Consequently, although qualified operators have to use more cognitive resources, the use of equally acceptable actions would be an inescapable choice to resolve this problem. As a result, Fig. 5.7 depicts additional action sequence about equally acceptable actions.

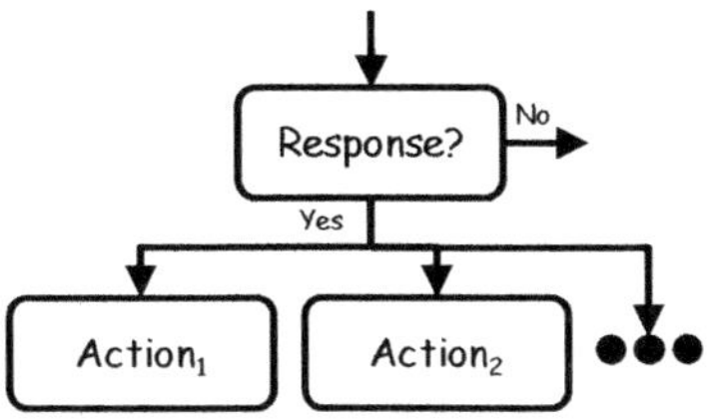

Type E

Fig. 5.7 Additional sequence of action – equally acceptable actions

Based on these definitions, it is possible to express the sequence of actions using a directed graph called an action control graph (ACG). For example, Fig. 5.8 depicts the ACG of the fourth procedural step shown in Fig. 5.5.

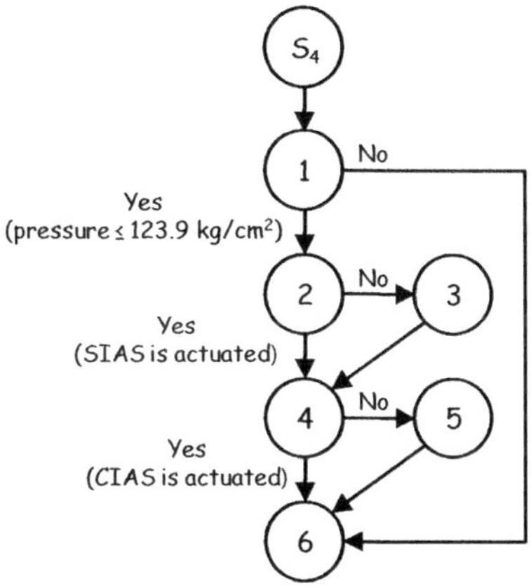

ID	Action description
S_4	Perform the fourth procedural step
1	Verify pressurizer pressure is less than 123.9 kg/cm^2
2	Verify SIAS is automatically actuated
3	Actuate SIAS manually
4	Verify CIAS is automatically actuated
5	Actuate CIAS manually
6	Go to the next procedural step

Fig. 5.8 ACG of the fourth procedural step shown in Fig. 5-5

From Fig. 5.8, it seems that an ACG is very similar to a software control flow graph of software, which is a directed graph with a unique start and end node. In addition, it appears that the ACG is very useful for visualizing the required actions with the associated sequence of actions that should be followed by qualified operators.

5.5 Performing Emergency Tasks

When an emergency event has occurred, most emergency tasks prescribed in EOPs are carried out by an operating team working in the main control room (MCR) of NPPs. Although there are several different types of team structures in NPPs (Moray 1999), Fig. 5.9 will be helpful to clarify the typical team structure of KSNPs with the associated responsibilities for the performance of emergency tasks.

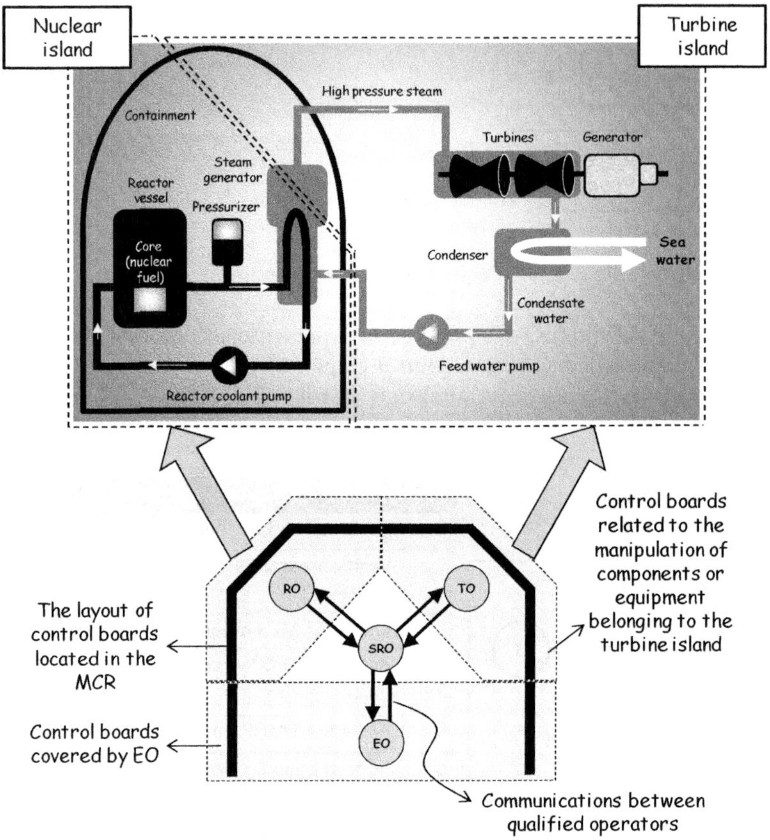

Fig. 5.9 The role of qualified operators working in the MCR of KSNPs

Each operating team working in the MCR of KSNPs consists of four qualified operators: (1) a senior reactor operator (SRO), (2) a reactor operator (RO), (3) a turbine operator (TO), and (4) an electrical operator (EO). In short, the SRO has the overall responsibility for the performance of emergency tasks, while the RO and the TO have a limited responsibility for the operation of components that belong to a nuclear island and a turbine island, respectively. Here, the nuclear island includes the primary circulation loop as well as all the components installed in a containment building. In contrast, the turbine island implies all the components included in the secondary as well as the third circulation loop. In addition, the EO simultaneously checks the status of electric power generation as well as the supplement of electrical power for all kinds of components installed in the nuclear and the turbine island.

Under this team structure, based on the SRO's command, each board operator (i.e., the RO, the TO, and the EO) has to manipulate many kinds of necessary components by using several control boards, in which many conventional control devices (such as alarm tiles, indicators, trend recorders or control devices, etc.) are located. In military parlance, this operation scheme is known as *the command and control operation.*

> The exercise of authority and direction by a properly designated commander over assigned and attached forces in the accomplishment of the mission. Command and control functions are performed through an arrangement of personnel, equipment, communications, facilities, and procedures employed by a commander in planning, directing, coordinating, and controlling forces and operations in the accomplishment of the mission (DOD 2009).

For example, let us consider a SRO who has to perform *Verify pressurizer pressure is less than 123.9 kg/cm^2* action. In order to perform this action, the SRO needs to know the current pressurizer pressure value. At this moment, the SRO tells the RO to read the current pressurizer pressure value because the pressurizer is one of the main systems in the nuclear island. Then the RO gives the SRO the desired information after reading the appropriate indicator. According to the RO's report, the SRO ultimately decides whether the pressurizer pressure is less than 123.9 kg/cm^2 or not. In this way, remaining required actions included in emergency tasks can be performed. More detailed information about the performance of emergency tasks can be found in Park et al. (2005).

References

AKIP (2008) Atomic Knowledge Information Portal. http://www.atomic.or.kr/ (in Korean)

CEOG (1996) Combustion engineering emergency response guidance. CEN-152, Rev. 04

Corcoran WR, Porter NJ, Church JF, Cross MT (1981) The critical safety functions and plant operation. Nuclear Technol 55:690–712

DOD (2009) DOD Dictionary of Military Terms. http://www.dtic.mil/doctrine/jel/doddict/data/c/01078.html

IAEA (1985) Developments in the preparation of operating procedures for emergency conditions of nuclear power plants. IAEA-TECDOC-341, Vienna

IAEA (1998) Good practices with respect to the development and use of nuclear power plant procedures. IAEA-TECDOC-1058, Vienna

Kadak AC, Candon JD (1984) A functional approach to transient management In: Lassahn PL, Majumdar D, Brockett GF (eds) Anticipated and Abnormal Plant Transients in Light Water Reactors, Plenum, New York, vol 2, pp.1127–1140

Korea Hydro and Nuclear Power (2002) Steam generator tube rupture. EOP-03, Younggwang

Macwan A, Mosleh A (1994) A methodology for modeling operator errors of commission in probabilistic risk assessment. Reliabil Eng Syst Saf 45:139–157

Meyer OR, Blackman HS, Ford RE, Naney LN (1987) Onsite assessment of the effectiveness and impacts of upgraded emergency operating procedures. NUREG/CR-4617, Washington, DC

Moray N (1999) Advanced displays, cultural stereotypes and organizational characteristics of a control room. In: Misumi J, Wilpert M, Miller R (eds) Nuclear Safety: A Human Factors Perspective. Taylor & Francis, New York

Nuclear Energy Institute (2008) http://www.nei.org/resourcesandstats/nuclear_statistics/

Nuclear Safety Information Center (2008) http://nsic.kins.re.kr/control/con_f00_01.asp?p_topMenuId=28&p_ME01_CODE_NUMX=446 (in Korean)

Park J, Jung W (2004) A study on the systematic framework to develop effective diagnosis procedures of nuclear power plants. Reliabil Eng Syst Saf 84(3):319–335

Park J, Jung W (2005) A database for human performance under simulated emergencies of nuclear power plants. Nuclear Eng Technol 37(5):491–502

Park SH, Kwon JS, Kim SR (1995) Emergency procedure recommendation for Wolsong 2, 3 & 4 NPP. In: Proceedings of the Korean Nuclear Society Autumn Meeting, Seoul, pp.272–277

Surman RC, Monty BS, Stella ME, Julian HV (1984) Guidance for control room emergency operations. In: Lassahn PL, Majumdar D, Brockett GF (eds) Anticipated and Abnormal Plant Transients in Light Water Reactors, Plenum, New York, vol 2, pp.1127–1140

USNRC (1982) Guidelines for the preparation of emergency operating procedures. NUREG-0899, Washington, DC

USNRC (2008) http://www.nrc.gov/reading-rm/doc-collections/fact-sheets/steam-gen.html

Virtual Nuclear Tourist (2008) http://www.virtualnucleartourist.com/

WOG (1987) Emergency response guidance, HP Volume, Rev. 1A

Wilkinson CD (1984) Elements of effective control room response to emergencies. In: Lassahn PL, Majumdar D, Brockett GF (eds) Anticipated and Abnormal Plant Transients in Light Water Reactors, Plenum, New York, vol 2, pp.1049–1057

6 Analyzing the Required Actions Prescribed in Emergency Tasks

As stated at the end of Chap. 4, the complexity of proceduralized tasks should be quantifiable by the concept of graph entropies if we construct a series of graphs representing the features of five kinds of complexity factors. In some respect, this requirement seems to be easily fulfilled because, for example, an ACG is directly comparable to the control flow graph of software. This implies that the effects of two kinds of complexity factors on the complexity of proceduralized tasks might be quantified from the ACG. That is, the first-order entropy of the ACG represents the contribution of logical entanglement on the complexity of proceduralized tasks, while the second-order entropy represents the contribution from the number of actions to be conducted by qualified operators. Unfortunately, we still need three more graphs that are able to characterize the remaining complexity factors: (1) the amount of information to be processed by qualified operators, (2) the amount of domain knowledge, and (3) the level of engineering decision.

Consequently, it may be necessary to meticulously analyze the contents of an action because the core of proceduralized tasks is to specify what should be done and how to do it. In other words, it is strongly expected that we can extract the necessary information to construct the corresponding graphs by scrutinizing the contents of an action. To clarify this aspect, let us look at the following explanations excerpted from Dougherty (1992).

> Potential rules in procedures, which we have generously assumed are candidates for rule-based behavior, include …
>
> 3. The symptom 'reactor level' in a BWR or 'subcooling margin' in a PWR.
> 4. The symptom 'Emergency Depressurization is Anticipated' in a BWR procedure.
>
> …
>
> The third case indicates that so-called symptoms may be simple instrumented parameters or more abstract or complex comparisons or interpretations. The fourth case is hard to analyze since the operant word is a human ability, anticipation, that may be used in variable, idiosyncratic ways by different people. Hence, it is hard to count the fourth item as an instruction at all (p. 254).

In other words, Dougherty criticizes the absence of essential contents of action descriptions, which results in the adoption of *variable and idiosyncratic ways* to accomplish proceduralized tasks (i.e., nonstandardized behaviors). Here, a departure from standardized behaviors implies the loss of an important benefit justifying why we have to use a procedure. Therefore, a systematic framework, by which the critical contents of an action are properly distinguished, should be determined first.

6.1 Key Contents of an Action Description

Existing works in the literature have stressed that there is a certain rule to write an effective action statement that directs what is needed to be done. For example, the Department of Energy (1998) made the following recommendation: "Complete the basic action step with supportive information about the action verb and the direct object. Supportive information includes further description of the object and the recipient of the object. Acceptance criteria, referencing, and branching are other types of supportive information … (p. 37)."

In addition, the Department of Defense (1999) explained that "The task inventory is composed of task statements, each of which consists of (a) an action verb that identifies what is to be accomplished in the task, (b) an object that identifies what is to be acted upon in the task, and (c) qualifying phrases needed to distinguish the task from related or similar tasks (p. 226)."

These recommendations give us an important clue to understand the key contents of an action description. That is, it is supposed that each action description can be decomposed into three parts: an *ACTION VERB*, an *OBJECT*, and an action specification. Since the meaning of *OBJECT* is self-explanatory (e.g., *a tangible and visible entity that is to be acted on*), it is worth focusing on the remaining parts.

6.1.1 Action Verb

Webster's New Millennium Dictionary of English defines an *ACTION VERB* as *a word belonging to the part of speech that is the center of the predicate and which describes an act or activity* (Webster 2008). In a technical term, the following definition seems to be more appropriate context: "A word that conveys action/behaviors and reflects the type of performance that is to occur (i.e., place, cut, drive, open, hold). Action verbs reflect behaviors that are measurable, observable, verifiable and reliable (Glossary 2008)."

This definition reflects the fact that an *ACTION VERB* is probably the most important part of describing an action. Subsequently, articulating *ACTION VERB*s should be the approach to characterizing actions to be performed by qualified operators. Table 6.1 summarizes the list of *ACTION VERB*s that has been commonly used in the EOPs of NPPs (DOE 1998; Jung 2001).

Table 6.1 Selected *ACTION VERB*s frequently appearing in EOPs

ID	ACTION VERB	Meaning
1	Align	Arrange equipment in a specific configuration to permit a specific operation
2	Close	Manipulate a device to allow the flow of electricity or to prevent the flow of fluids, other materials, or light
3	Cool (down)	Lower the temperature of equipment or environment
4	Depressurize	Release gas or fluid pressure
5	Determine	Find out; ascertain
6	Energize	Provide equipment with electrical power
7	Ensure	Confirm that an activity or condition has occurred in conformation with specific requirements
8	Increase	Produce a larger value
9	Isolate	Shut off or remove from service
10	Maintain	Hold or keep in any particular state or condition, especially in a state of efficiency or validity
11	Open	Manipulate a device to prevent the flow of electricity or to allow the flow of fluids, other materials, or light
12	Operate	Cause equipment or system to perform designated functions
13	Perform	Carry out specified actions
14	Reduce	Decrease a variable to meet a procedure requirement
15	Reset	Restore a piece of equipment, part, or component to a previous condition, parameter value, instrument set point, or mechanical position
16	Stabilize	Become stable, firm, steady
17	Start	Initiate the function of an electrical or mechanical device
18	Stop	Halt movement or progress; hold back
19	Throttle	Adjust a valve to an intermediate position to obtain a desired parameter value
20	Verify	Confirm, substantiate, and assure that a specific activity has occurred or that a stated condition exists

6.1.2 Action Specification

The next part is an action specification that might be *supportive information* that helps qualified operators to carry out an action or *qualifying phrases* needed to distinguish each action from related and/or similar actions. For example, let us recall the following two actions pertaining to making a smooth cookie batter, which are exemplified in Sect. 1.3.

A1 Cream together the butter and the brown sugar until smooth
A3 Using a mixer fitted with paddle attachment, cream butter and sugar

together until very light, about 5 min

Here, we can decompose the key contents of these actions into three parts as shown in Table 6.2.

Table 6.2 Comparing key contents of two arbitrary actions

Action description	Contents	Corresponding description
Cream together the butter and the brown sugar until smooth	ACTION VERB	Cream
	OBJECT	Batter (mixture of butter and sugar)
	Action specification	Until smooth
Using a mixer fitted with paddle attachment, cream butter and sugar together until very light, about 5 min	ACTION VERB	Cream
	OBJECT	Batter
	Action specification	• Until very light • A mixer with a paddle (a dedicated means) • Operation time (5 min)

From Table 6.2, it is evident that two actions share the same *ACTION VERB* as well as *OBJECT*. However, although action A3 is lengthier, it is expected that this action will be accomplished more easily than action A1. One plausible reason is the difference in the action specification. That is, the action specification of the former is quite subjective (i.e., until smooth) while that of the latter is objective (i.e., specifying how long the mixer is to be used). As a consequence, it is anticipated that the former action will require more cognitive resources to decide whether the batter is smooth or not.

In fact, Bovair and Kieras (1996) cited the result of a previous study pertaining to writing an effective procedural instruction:

> They found that the good and bad instructions could not be distinguished by text characteristics likely to affect reading comprehension such as length of text or length of sentences; in deed, some of the best instructions had the most complex syntax and sentence structure. The important differences between good and bad instructions seemed to be those of contents; in particular, poor instructions omitted important details like the orientation of parts in the assembly task, and often included the wrong level of detail (p. 222).

This result strongly indicates that even if the length of an action description becomes longer, it is much more important for qualified operators to provide appropriate action specifications. Conversely, if qualified operators feel any burden in performing an action, it is assumed that this burden would have been largely caused by insufficient action specifications. This means that analyzing the characteristics of action specifications will give an important clue in identifying the contents that should be included in an action.

6.2 Characterizing an Action

In order to identify the characteristics of action specifications, detailed analysis has been carried out for all the EOPs of KSNPs (Park *et al.* 2005). As a result, three radical elements related to action specifications and two types of peculiarities have been distinguished as summarized in Table 6.3.

Table 6.3 Characterizing scheme of actions included in EOPs

Category	Element	Predefined property
Action specification	MEANS	Designated means (DEG)
		Inherent means (INH)
		No means (NM)
		Local operation (LO)
	ACCEPTANCE CRITERION	Objective criterion (OBJ)
		Reference information (RI)
		Subjective criterion (SUB)
		No criterion (NC)
	CONSTRAINT	Objective constraint (OBJ_C)
		Subjective constraint (SUB_C)
		Reference information (RI_C)
		No limitation (NL)
Peculiarity	Selection (SEL)	Yes or No
	Continuous control (CC)	Yes or No

6.2.1 Means

A *MEANS* indicates an explicit method that specifies how to achieve the expected state of a given action. The *MEANS has four properties*: (1) designated means (*DEG*), (2) inherent means (*INH*), (3) no means (*NM*), and (4) local operation (*LO*). For example, let us compare the following three actions:

- Cool down the temperature of the RCS to 275°C using valve A
- Close valve A
- Cool down the temperature of the RCS to 270°C

It is evident that the goal of the first action is to cool down (*ACTION VERB*) the temperature of thc RCS (*OBJECT*). To accomplish this goal, this action forces qualified operators to use the value A. In other words, even though other valves are available to reduce the temperature of the RCS, this action must be accomplished by manipulating the valve A. Therefore, *DEG* is assigned to the first action.

Meanwhile, the second action did not specify any method to close the valve A. However, the omission of a specific method seems to be acceptable, if it is assumed that the only way to close the valve A (i.e., the goal of this action) is to use the associated controller (i.e., the controller of the valve A). In other words, although there is no specification about *MEANS*, it is assumed that the action already implies the proper method if there is no choice for accomplishing its goal. Accordingly, in order to distinguish the term of *DEG*, the second action is regarded as an action that contains *INH*.

Similarly, the third action does not prescribe any specific method to lower the temperature of the RCS. However, the implication of this omission is completely different from that of the second action, because it is assumed that there are several equivalent methods to reduce the temperature of the RCS. This indicates that *NM* should be assigned to the third action, because qualified operators have to come up with an appropriate method to lower the temperature of the RCS.

It is to be noted that *NM* should be assigned to an action that does not manifest the associated components or equipment requiring the intervention of qualified operators. For instance, *NM* should be assigned to the action *align all the valves to transfer a coolant from Tank A to Tank B* because it does not specify the associated valves that are necessary to make a flow line from *Tank A* to *Tank B*.

As for the last, it is necessary to clarify the meaning of *LO*. Let us assume an arbitrary action, such as *ensure that a field operator stopped pump C*. In this case, it is obvious that the purpose of this action is to verify whether a field operator who is working in a local place successfully stopped the *pump C* or not. In this case, it would be difficult to determine which controller will be used, because the field operator is liable to select the most appropriate one available in that particular location. Therefore, in order to distinguish *NM* as well as *INH*, *LO* should be assigned to an action requiring the assistance or cooperation of field operators working at that location.

6.2.2 Acceptance Criterion

It is apparent that there are many actions requiring the decision of qualified operators, such as *verify SIAS is automatically actuated*. Accordingly, an *ACCEPTANCE CRITERION*, by which qualified operators confirm whether the goal of a given action is achieved or not, should be regarded as the important element of action specifications (DOE 1998).

In many cases, the *ACCEPTANCE CRITERION* articulates either the state that an *OBJECT* is expected to reach or any condition by which the current status of an *OBJECT* can be confirmed. Thus, we can consider four kinds of properties in characterizing the *ACCEPTANCE CRITERION*: (1) objective criterion (*OBJ*), (2) reference information (*RI*), (3) subjective criterion (*SUB*), and (4) no criterion (*NC*).

First, let us recall *close valve A* action whose expected status is a fully closed valve position. Therefore, the success or failure of this action can be easily determined by checking a valve status indicator. Similarly, in the case of *verify pressurizer pressure is less than 123.9 kg/cm^2* action, qualified operators can confirm whether the current status has reached the expected status or not because there is a clear *ACCEPTANCE CRITERION* – less than 123.9 kg/cm^2. Therefore, any *ACCEPTANCE CRITERION* that provides an unbiased yardstick is regarded as *OBJ*. Table 6.4 summarizes typical examples of *OBJ*.

Table 6.4 Several examples of *OBJ*

Property	Example	The associated action
Dichotomous	Open/Close	Close main feed water isolation valves (MFIVs)
	On/Off	Verify safety injection actuation signal (SIAS) is actuated
	Start/Stop	Stop all RCPs
Discrete value	$\geq$ (greater than)	Verify subcooling margin is greater than 15°C
	$\leq$ (less than)	Verify pressurizer pressure is less than 123.9 kg/cm^2
Explicit range	135~165 kg/cm^2	Verify pressurizer pressure is maintained within 135~165 kg/cm^2
Trend	Increase	Verify pressurizer pressure is increasing
	Decrease	Verify pressurizer pressure is decreasing

Second, although the *ACCEPTANCE CRITERION* is manifested in the required action, there are times when qualified operators are not able to directly apply it. For example, let us consider an action, such as *verify sufficient safety injection (SI) flow is delivered to RCS (refer to SI delivery curve)*, whose goal is to confirm the delivery of a *sufficient* SI flow. Here, should to be noted that the satisfaction of the expected state should be determined by a reference curve like Fig. 6.1.

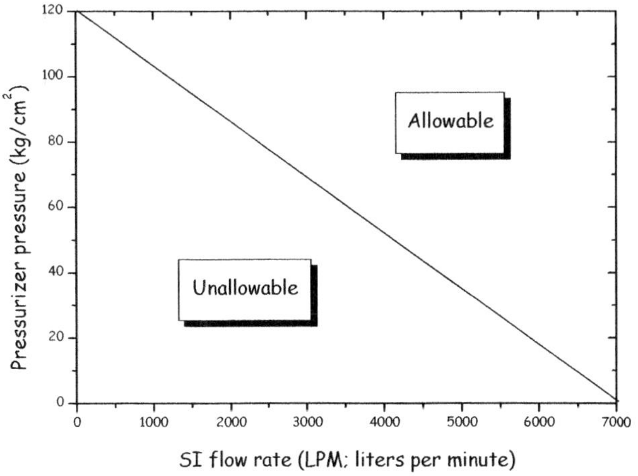

Fig. 6.1 Hypothetical curve to determine the delivery of a sufficient SI flow

In Fig. 6.1, in order to confirm the delivery of a sufficient SI flow (i.e., *acceptable* area), qualified operators have to compare the current SI flow rate with the expected rate that varies with respect to pressurizer pressure. This implies that qualified operators need to confirm the satisfaction of an *ACCEPTANCE CRITERION*, not by the simple observation of an associated indicator but by the integration of additional information to identify the status of an ongoing situation. For this reason, *RI* is considered one of the properties of the *ACCEPTANCE CRITERION*. Table 6.5 shows several actions whose acceptance criterion can be confirmed by *RI*.

Table 6.5 Properties of *RI* with the associated actions

Property	Meaning	Associated action
Time	Reference information is given by a certain period of time	Verify feed flow has been supplied for the preceding 5 min
Figure/Chart	Reference information is given by figures or charts	Verify sufficient SI flow is delivered to RCS (refer to SI delivery curve)
Table/List	Reference information is given by tables or lists	Cool down the temperature of the ruptured SG to a target temperature (refer to *Table X*)
Equation/Formula	Reference information can be obtained from equations or formulas	Determine the leak rate of an isolation valve (refer to *Equation Y*)
Static configuration	The information about component configurations is used as reference information	Close isolation valve linked to the discharge line (i.e., a valve linked to the discharge line can be determined by static configuration)
Dynamic configuration	Component configurations that vary due to an ongoing situation are regarded as reference information	Isolate auxiliary feed water flow delivered to the ruptured SG (i.e., the ruptured SG dynamically varies with respect to the location of ruptured tubes)

Third, there are times when qualified operators suffer from an ambiguous *ACCEPTANCE CRITERION*. For example, let us consider *verify pressurizer pressure is abnormally decreasing* action. Unfortunately, qualified operators will likely make different decisions when they are faced with this action. This is probably because the subjectivity (or ambiguity) of the *ACCEPTANCE CRITERION*, which forces qualified operators to make a tricky decision – *which tendency represents abnormally decreasing pressurizer pressure*? or *how can we confirm the decrease of pressurizer pressure is not a natural phenomenon in this situation*? Accordingly, an *ACCEPTANCE CRITERION* that is able to provide a biased yardstick is referred to as *SUB*. Table 6.6 shows typical examples.

However, the worst case is an action that does not have any *ACCEPTANCE CRITERION*. In this case, as with the last property, *NC* is assigned to the action. For example, *NC* should be assigned to *stabilize pressurizer pressure using pres-*

surizer spray valves action because this action consists of *ACTION VERB* (stabilize), *OBJECT* (pressurizer pressure), and *MEANS* (pressurizer spray valves) without any specification about the *ACCEPTANCE CRITERION* (i.e., how to define the status of a *stabilized* pressure).

Table 6.6 Typical examples of *SUB*

Property	Example	Associated action
Status	Uncontrollable (or controllable)	Verify there is no SG whose pressure is decreasing in an uncontrolled manner
	Abnormal (or normal)	Verify pressurizer pressure is abnormally decreasing
	Unstable (or stable)	Ensure the pressure of each SG is stable
Potentiality	The possibility of restoration	Determine that at least one AC (alternating current) emergency bus can be restored
	Necessity (or anticipation)	Open supply breakers for all unnecessary DC (direct current) loads

6.2.3 *Constraint*

A *CONSTRAINT* represents a restriction (or limitation) that has to be obeyed to accomplish the goal of a given action. At a glance, the purpose of the *CONSTRAINT* seems to be similar to that of an *ACCEPTANCE CRITERION*, because they commonly deal with a condition to be satisfied. This implies that the identical set of properties considered in the *ACCEPTANCE CRITERION* can be applied to the *CONSTRAINT*. That is, the *CONSTRAINT* has four kinds of properties: (1) objective constraint (*OBJ_C*), (2) reference information (*RI_C*), (3) subjective constraint (*SUB_C*), and (4) no limitation (*NL*).

However, it should be noted that there is a difference between the *ACCEPTANCE CRITERION* and the *CONSTRAINT* because the former specifies the expected (or final) status of an *OBJECT* while the latter clarifies a condition related to an *ACTION VERB* or a *MEANS*. For example, let us consider *open steam bypass control system (SBCS) valve #1 to 100%, until RCS temperature is less than 260°C* action. In this action, the *ACTION VERB*, *OBJECT*, and *ACCEPTANCE CRITERION* are *open*, *SBCS valve #1*, and *100%*, respectively. However, the phrase starting with *until* denotes an additional condition that fixes when qualified operators have to close SBCS valve #1 (i.e., OBJ_C). Similarly, the *CONSTRAINT* of *close feed water control valve #1 when SG level becomes stable* action is *SUB_C* because it subjectively defines the timing for when qualified operators have to close feed water control valve #1 (e.g., the interpretation of a *stable* SG level would be subjective).

6.2.4 *Peculiarity*

In characterizing an action, the aforementioned elements are identified from the point of view of action specifications. In addition to this, it is indispensable to consider a *peculiarity* that pertains to the performance of an action. It is to be noted that, although there would be more peculiarities, two types of peculiarities are considered in this book. The first one is related to the selection of an action. Let us look at the following procedural step containing equally acceptable actions.

IF necessary, perform **ANY** of the following.

- Stop HPSI (high pressure safety injection) pumps
- Throttle HPSI flow
- Operate PLCS (pressurizer level control system)
- Operate charging pumps

From the point of view of action specifications, Table 6.7 shows the result of decompositions of the first two actions.

Table 6.7 Action descriptions, elements, and their properties with respect to equally acceptable actions

Action description	Element	Property
If necessary, perform any of the following	OBJECT	Any of the following
	MEANS	NM
	ACCEPTANCE CRITERION	SUB (necessity)
	CONSTRAINT	NL
Stop HPSI pumps	OBJECT	HPSI pumps
	MEANS	INH
	ACCEPTANCE CRITERION	OBJ (dichotomous)
	CONSTRAINT	NL

In Table 6.7, it is observed that there is a problem in characterizing the first action. That is, from the point of view of action specifications, the first action seems to be very unusual because its *OBJECT* does not clarify a tangible and visible entity, such as HPSI pumps. Meanwhile, this action forces qualified operators to select an appropriate *OBJECT* (i.e., any action must be done). Therefore, to resolve this problem, it would be better to define another property by which the nature of the selection can be represented. As a result, instead of considering five actions, the above procedural step is regarded as a procedural step that consists of four actions with the *peculiarity* of *SEL* (Table 6.8).

Another *peculiarity* is related to an action that requires the continuous control activity of qualified operators. A typical example is an action that forces qualified operators to adjust a process parameter, such as *cool down the temperature of RCS to 275°C using valve A*. To accomplish this action, qualified operators should continuously adjust the open position of the associated valve as well as monitor the RCS temperature until the target temperature is reached. Therefore, this action seems to be very unique, because it impels qualified operators to continuously use their cognitive resources for an extended period. For this reason, it is necessary to distinguish actions requiring a continuous control activity from other actions by assigning them the designation *CC*.

Table 6.8 A set of actions that are interlinked by SEL property

Action description	Element/Peculiarity	Property
Stop HPSI pumps	OBJECT	HPSI pumps
	MEANS	INH
	ACCEPTANCE CRITERION	OBJ (dichotomous)
	CONSTRAINT	NL
	Peculiarity	SEL
Throttle HPSI flow	OBJECT	HPSI flow
	MEANS	NM
	ACCEPTANCE CRITERION	NC
	CONSTRAINT	NL
	Peculiarity	SEL
Operate PLCS	OBJECT	PLCS
	MEANS	INH
	ACCEPTANCE CRITERION	OBJ (dichotomous)
	CONSTRAINT	NL
	Peculiarity	SEL
Operate charging pumps	OBJECT	Charging pumps
	MEANS	INH
	ACCEPTANCE CRITERION	OBJ (dichotomous)
	CONSTRAINT	NL
	Peculiarity	SEL

6.3 Constructing Graphs

Based on the result of action decompositions as presented in Table 6.8, we are able to construct a set of graphs that characterize three kinds of complexity factors: (1) the amount of information to be processed by qualified operators, (2) the amount of domain knowledge that is indispensable to perform the required action, and (3) the level of engineering decision related to the establishment of an appropriate decision criterion to perform the required actions.

6.3.1 Information Structure Graph

The first graph we need to construct is one that is able to characterize the requisite information to accomplish the required actions. In other words, the amount of information to be processed by qualified operators should be represented by this graph. To this end, it is necessary to answer a preliminary question: *What kind of information should be managed to perform proceduralized tasks*? In connection with this question, it is to be noted that most qualified operators working in the MCR of PWRs have performed emergency tasks by using conventional control devices, such as push buttons, knobs, indicators, measurements about process parameters, trend recorders, and alarm tiles, etc. This means that the information to be managed by qualified operators can be expressed by the combination of five types of basic information shown in Table 6.9 (Lee *et al.* 2008).

Table 6.9 Basic information types in a conventional MCR

Basic type	Meaning	Canonical example
Boolean (B)	Qualified operators need to manage binary information	Identifying the existence of process alarms
	Qualified operators need to manipulate a component that has a binary operating mode	Manipulating a valve (open/close) or a pump (start/stop), etc.
Float (F)	Qualified operators need to manage the value of a process parameter presented by a real number	Reading pressure, temperature, flow rate, etc.
Integer (I)	Qualified operators need to manage the value of a process parameter presented by an integer	Indentifying the number of cooling fans under operation
Array of Boolean (AB)	Qualified operators need to manipulate a component that has several kinds of operating modes	Manipulating a valve or a pump having several operating modes, such as open, close, auto, etc.
Array of Float (AF)	Qualified operators need to manipulate a component that can be continuously adjusted	Manipulating a valve that can continuously adjust its open position
	Qualified operators need to determine the trend of a process parameter	Identifying the trend (increase, decrease, constant) of pressure, temperature, flow rate, etc.

In addition, it is believed that qualified operators can accomplish the required action more easily and correctly when they are given critical information compatible with the three radical elements of action specifications. As an example, let us consider *stop HPSI pumps* action. In this case, although there is no detailed description, qualified operators would be expected to already know appropriate controllers to stop HPSI pumps (i.e., *INH*). In addition, since there is no *CONTRAINT* in this action, qualified operators need to access information by which the stoppage of HPSI pumps can be directly confirmed. This implies that qualified operators have to manage at least two kinds of information related to (1) the manipulation of HPSI pump controllers (*MEANS*) and (2) the confirmation of desired states (*ACCEPTANCE CRITERION*). Accordingly, it is possible to construct the information structure graph (ISG) of this action, which corresponds to the data structure graph of software (Fig. 4.2). To clarify this aspect, it will be helpful to compare two kinds of arbitrary control environments as depicted in Fig. 6.2.

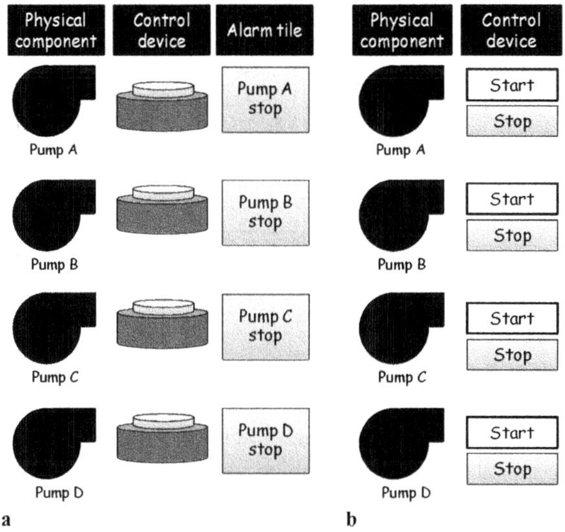

Fig. 6.2 Two kinds of arbitrary control environments

In Fig. 6.2a, the manipulation of each pump can be done by a push button that only has two operating modes (or functions), such as start or stop. In addition, there are four alarm tiles dedicated to informing qualified operators of the status of the pumps. In contrast, Fig. 6.2b shows four selection buttons that allow qualified operators not only to control the pumps but also to see their status. In other words, since a selected operating mode can be highlighted by a different color or blinking light, qualified operators easily identify the status of the pumps without accessing other sources of information (Lee et al. 2008).

Accordingly, even though qualified operators perform the same actions, different ISGs can be constructed due to the difference in control environments. That is, qualified operators who have to stop pumps in a control environment like that shown in Fig. 6.2a need to simultaneously manage two kinds of information, while those working in a control environment like that shown in Fig. 6.2b can accomplish the required action with a single source of information. Figure 6.3 shows two kinds of ISG that represent the amount of information to be managed by qualified operators.

It is not surprising that there are many required actions with the same source of information for a *MEANS* and an *ACCEPTANCE CRITERION*. For example, let us consider *verify pressurizer pressure is less than 123.9kg/cm^2* action.

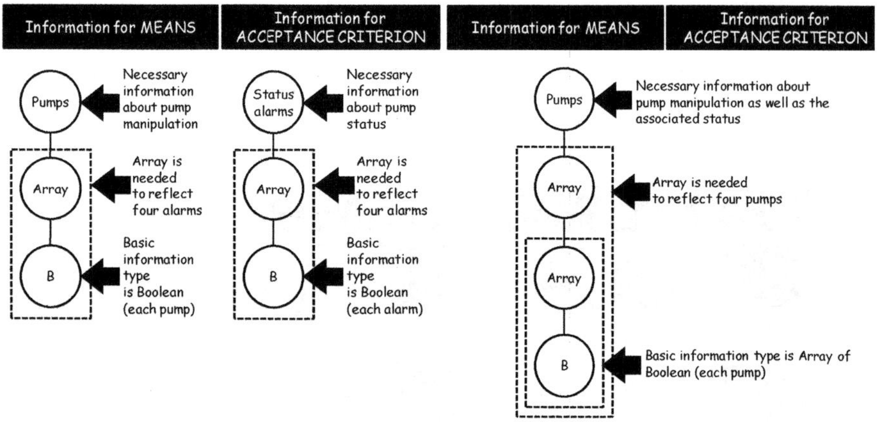

ISG for Fig. 6.2a **ISG for Fig. 6.2b**

Fig. 6.3 Two kinds of ISG due to different control environments

In this case, although there is no description about the MEANS, it seems to be evident that qualified operators should access the pressurizer pressure indicator (i.e., *INH*). In addition, in order to determine whether the *ACCEPTANCE CRITERION* of this action is satisfied or not, qualified operators need to read the current pressurizer pressure value. This implies that the source of necessary information pertaining to the *ACCEPTANCE CRITERION* is also the pressurizer pressure indicator. Accordingly, the ISG of this action can be depicted as in Fig. 6.4.

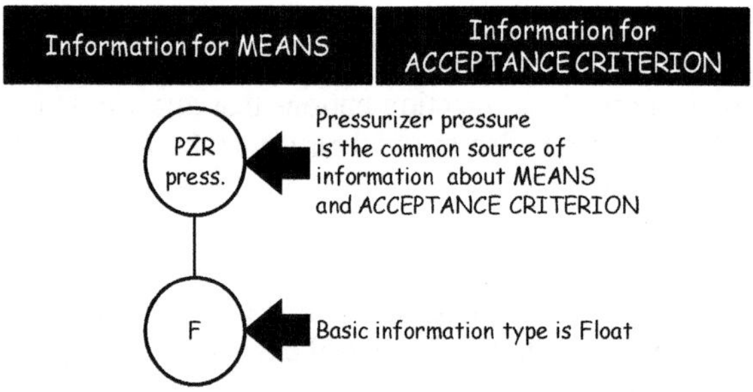

Fig. 6.4 ISG of an action that shares the same source of information about *MEANS* and *ACCEPTANCE CRITERION*

6.3.2 *Abstraction Hierarchy Graph*

Next, we have to construct a graph that can determine the amount of domain knowledge needed to perform the required action. In this regard, Moray (1998) pointed out that qualified operators usually accumulate domain knowledge in a hierarchical way.

> Thus, an operator may initially learn all the details of the controls in a control panel, but later come to think of them not as 'Valve 1, Valve 2, Pump 6,' etc., but as 'Cooling system,' Steam generator,' etc. This description in turn can be remodeled into 'Power Generation, Power distribution,' etc. Thus, operators construct a hierarchical set of models as a series of many-to-one mappings (p. 295).

In other words, qualified operators should start to build their domain knowledge from the component level to a higher level that consists of several components. In addition, over time, qualified operators will repeat the integration of a lower level knowledge in order to get a higher level knowledge. This strongly suggests that the amount of domain knowledge can be represented in the form of a hierarchical graph that is very similar to a software data structure graph.

With this in mind, we are able to adopt the framework of an abstraction hierarchy (AH), which was developed under the context of a supervisory control task (Rasmussen 1986). According to the AH framework, any human-made physical system can be analyzed by the following five levels of inherent functions.

- *Functional purpose* The intended functional effect of a system on its environment, such as the generation of electricity for NPPs.
- *Abstract function* The overall function of a system, which is represented by a causal structure such as mass or energy.
- *Generalized function* A set of basic functions that represent the functional structure of a system above the level of standard components.
- *Physical function* The characteristics of standard components, which can be clearly distinguished from their intrinsic functions, such as the function of pumps or valves, etc.
- *Physical form* The physical appearance of a component, such as its shape, weight, color, etc.

Based on these definitions, Rasmussen (1976, 1986) and Vicente (1999) emphasized that the AH framework is a remarkable tool for extracting the characteristics of domain knowledge to be considered by qualified operators. For this reason, it is expected that the AH framework can be used as a theoretical basis to identify the level of domain knowledge. Accordingly, as summarized in Table 6.10, four levels of domain knowledge are defined based on the results of a previous study (Jung 2001).

Table 6.10 shows that there are three differences between Rasmussen's AH framework and the four levels of domain knowledge. The first one is that domain knowledge corresponding to the *physical form* of the AH framework is excluded from the classification of domain knowledge because it was assumed that qualified operators would carry out proceduralized tasks. In other words, as stated in Sect. 3.2, since qualified operators have a minimum level of domain knowledge, it is believed that they would already have domain knowledge about the physical form of a component.

Table 6.10 Four levels of domain knowledge

Rasmussen's AH	Level of domain knowledge	Meaning
Abstract function	Abstract function (AF) related domain knowledge	Qualified operators need domain knowledge for delineating mass or energy flow based on two or more process functions or conditions
	Process function (PF) related domain knowledge	Qualified operators need domain knowledge for describing mass or energy flow based on two or more system functions or conditions
Generalized function	System function (SF) related domain knowledge	Qualified operators need domain knowledge that is related to two or more component functions or conditions
Physical function	Component function (CF) related domain knowledge	Qualified operators need domain knowledge that is related to the condition or function of a component, such as a valve, pump, heat exchanger and a heater, etc.

The second difference is the exclusion of domain knowledge related to the *functional purpose* defined in the AH framework. That is, it is futile to describe the required actions at the level of the functional purpose because such actions should provide qualified operators with detailed action specifications. In other words, *minimize the release of a radioactive material into the environment* action that describes one of the ultimate goals of EOPs is not helpful in providing detailed actions that qualified operators really want to know – *what is to be done* or *how to do it.*

The last difference is that domain knowledge pertaining to the *abstract function* of the AH framework has been subdivided into two levels, such as the *abstract function* and *process function* related domain knowledge. For example, let us consider two arbitrary actions: (1) *maintain core heat removal* and (2) *maintain the primary circulation.* According to the AH framework, both actions must belong to the abstract function level because they deal with the overall functions pertaining to the balance of mass and energy flow of PWRs (Sect. 5.1). However, these two actions seem to be distinguishable because the latter would be a subset of the former (i.e., one plausible way of maintaining core heat removal is to maintain the primary circulation). This strongly implies that qualified operators may need broader domain knowledge when the former action is called for. Therefore, to resolve this problem, the process function is introduced in Table 6.10. As a result, Fig. 6.5 shows an abstraction hierarchy graph (AHG) that can be used to represent the amount of domain knowledge needed by qualified operators.

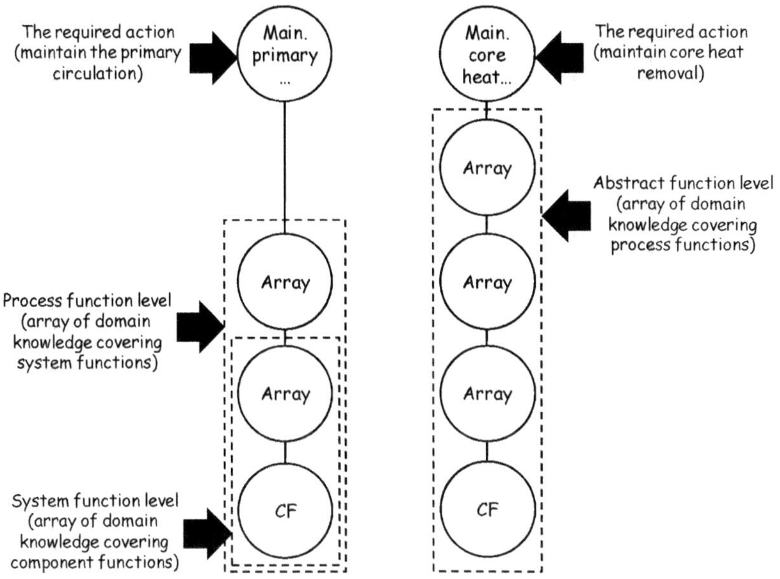

Fig. 6.5 AHGs of two arbitrary actions

6.3.3 Engineering Decision Graph

The last graph that we have to construct is related to the level of an engineering decision by which the amount of cognitive resources needed to establish the decision criteria of the required actions can be expressed. In this regard, although there is no explicit rule, it is assumed that qualified operators usually accomplish a task demanding a high-level cognitive activity by decomposing it into a series of subtasks demanding lower-level cognitive activities (Rasmussen 1976; Hollnagel 1993a). For example, Ullman and D'Ambrosio (1995) found that engineers decompose design problems into manageable subproblems. In addition, Shugan (1980) pointed out that the cost of thinking could be captured by a measurable (i.e., well-defined and calculable) unit of thought, such as the average cost per binary comparison. Similarly, Jiang and Klein (2000), Johnson and Payne (1985), Spence and Tsai (1997), and Todd and Benbasat (2000) commonly stated that any cognitive process can be represented by a sequence of elementary cognitive activities or skills, such as comparing and recalling, etc.

The above rationales strongly support the idea that the decomposition of a complicated task is practical problem-solving technique. Actually, Bainbridge (1997) asserted that "For example, the task goal "keep temperature 300°C," involves the cognitive goals "find current temperature," "evaluate actual against required temperature," "choose corrective action." These steps may not be consciously explicit or distinct to the person doing the task (p. 355)."

This indicates that the level of engineering decision can be represented in the form of a hierarchical graph, in which the required action demanding a higher-level engineering decision is regarded as a series of actions demanding lower-level engineering decisions. To this end, it is indispensable to establish a technical basis by which the level of the engineering decision can be properly distinguished. In light of this concern, it would be very helpful to introduce a decision ladder model developed by Rasmussen (1974) because it depicts the decision making process of qualified operators who are dealing with a supervisory control task. Figure 6.6 shows the overall structure of the decision ladder model.

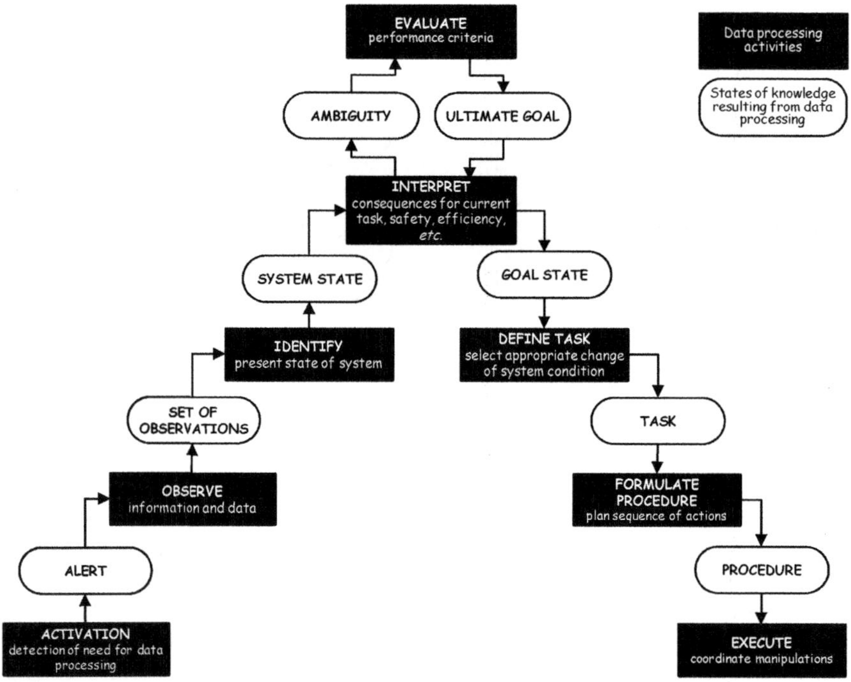

Fig. 6.6 The decision ladder model (see p. 27 of Rasmussen 1974)

Here, it should be noted that we need to be aware that the decision ladder model needs to be simplified when there is a procedure to be followed by qualified operators because of two reasons. First, since qualified operators already know what needs to be done, the *ACTIVATION* activity (i.e., detection of need for data processing) is less meaningful. Second, in most cases, qualified operators do not need to formulate the sequence of actions by themselves, because proceduralized tasks already have a predefined sequence of actions. As a result, Fig. 6.7 illustrates the simplified version of the decision ladder model.

From the simplified decision ladder model, it is possible to classify four levels of engineering decision pertaining to the performance of proceduralized tasks. To this end, let us consider an arbitrary system depicted in Fig. 3.2 with four arbitrary actions listed in Table 6.11.

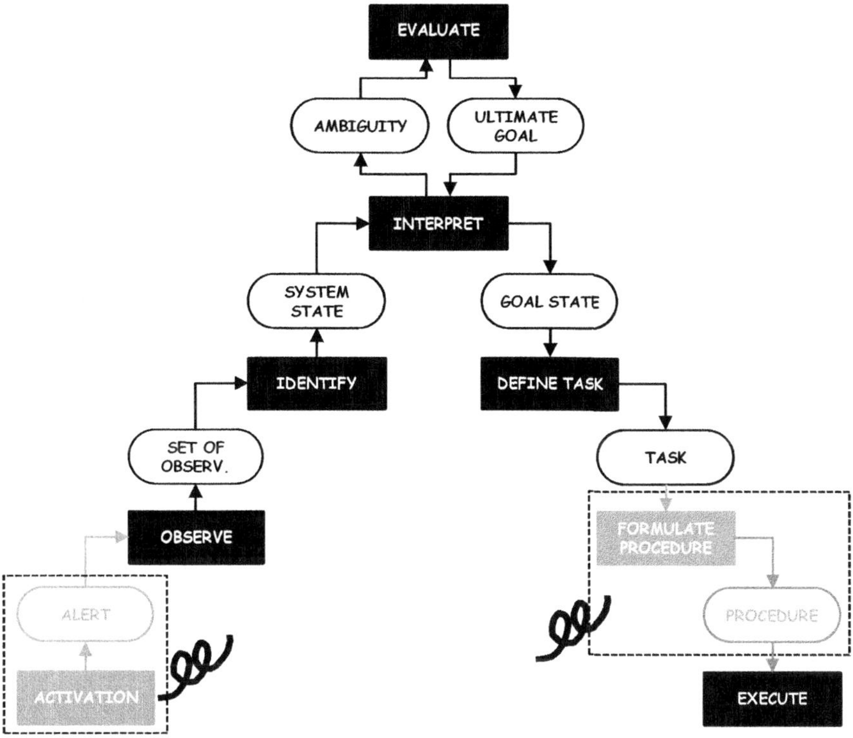

Fig. 6.7 Simplified decision ladder model to deal with a special situation in which qualified operators have to follow proceduralized tasks

Table 6.11 Four arbitrary actions to explain the levels of the engineering decision

ID	Action description
1	Verify the water level of Tank 1 is less than 30%
2	Verify the water level of Tank 1 is decreasing
3	Verify the water level of Tank 1 is abnormally decreasing
4	If necessary, perform any of the following. • Increase outflow • Provide bypass line

First of all, when qualified operators faced with *verify the water level of Tank 1 is less than 30%* action, they will start this action by observing the water level of Tank 1, because this is probably essential information to decide whether the water level of Tank 1 is less than 30% or not. Then, qualified operators will make a decision by comparing the observed water level with the *ACCEPTANCE CRITERION* of this action. From the point of view of the decision ladder model, a plausible sequence could be illustrated as Fig. 6.8.

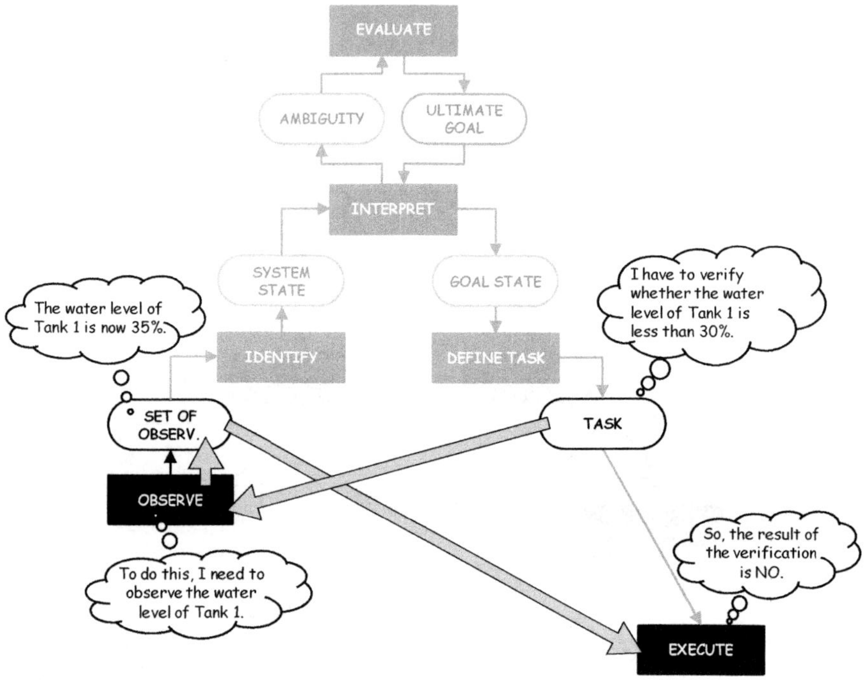

Fig. 6.8 Example explaining the sequence of decision making activities when qualified operators need to carry out *verify the water level of Tank 1 is less than 30%* action

Second, in order to perform *verify the water level of Tank 1 is decreasing* action, qualified operators will start this action by observing the water level of Tank 1. They will also realize that they have to keep observing the water level of Tank 1 for a while (i.e., collecting data about the water level with respect to time). Based on the collected data, they will identify the state of Tank 1, then they will finally make a decision about whether the water level is falling off or not. Figure 6.9 represents the plausible sequence of the associated decision making activities based on the simplified decision ladder model.

As shown in Fig. 6.9, it is expected that qualified operators will identify the state of Tank 1 by a set of data related to the changes of the water level over a time. This indicates that, as mentioned at the beginning of this section, it is possible to think of a state identification as a combination of lower-level cognitive activities, such as repeated *OBSERVE* activities. For this reason, a symbol signifying a circulation is inserted in the *OBSERVE* activity in Fig. 6.9.

Third, if qualified operators have to perform *verify the water level of Tank 1 is abnormally decreasing* action, then they will carry out a series of decision making activities that are similar to those related to *verify the water level of Tank 1 is decreasing* action. However, it is assumed that qualified operators will have to additionally perform the *INTERPRET* activity as illustrated in Fig. 6.10.

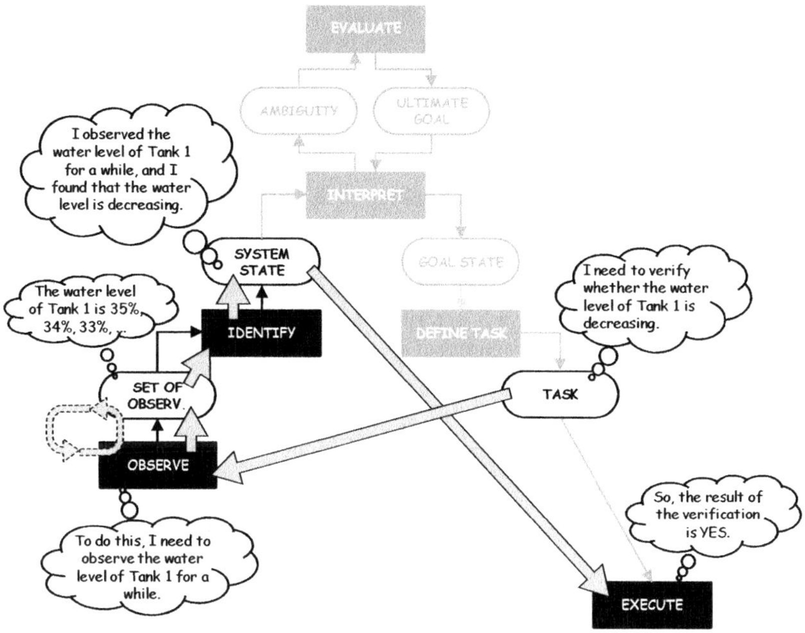

Fig. 6.9 Example of the sequence of cognitive activities pertaining to *verify the water level of Tank 1 is decreasing* action

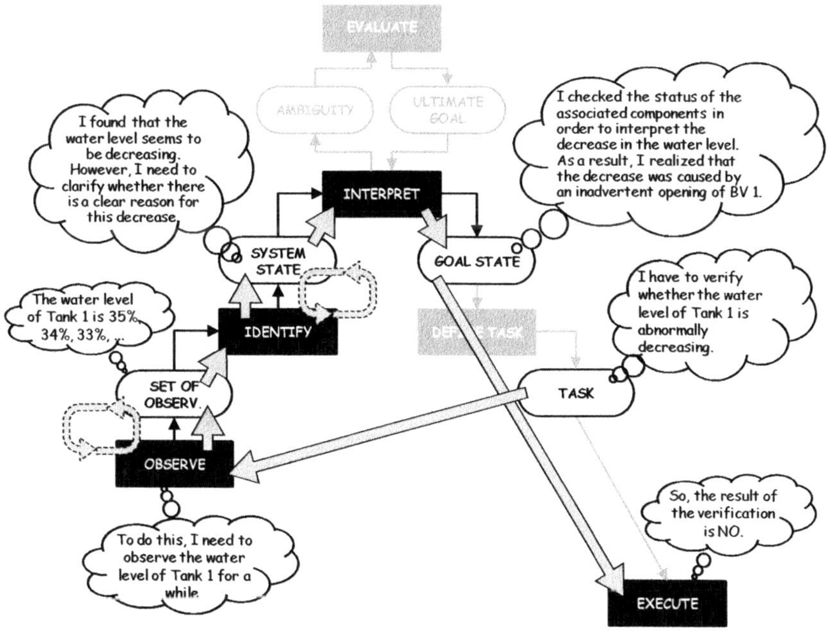

Fig. 6.10 Example of the sequence of decision making activities related to *verify the water level of Tank 1 is abnormally decreasing* action

As can be seen from Fig. 6.10, when the water level of Tank 1 seems to be decreasing, qualified operators have to decide whether or not this tendency can be explained. In other words, if there is a clear reason why the water level is decreasing, this symptom would be regarded as a normal response. In contrast, if there is no probable cause, then qualified operators will decide that the water level of Tank 1 is abnormally decreasing due to other reasons, such as a break in a pipe.

In order to make this kind of determination, qualified operators may repeatedly collect supplementary information such as the status of components that are able to cause a decrease in the water level of Tank 1 (e.g., the state of BV 1 as well as BV 2 or the open position of CV 1). For this reason, a symbol signifying circulation is inserted in the *IDENTIFY* activity. In addition, it is assumed that the *INTERPRET* activity could be expressed as a series of lower-level cognitive activities (i.e., the repetition of *IDENTIFY* as well as *OBSERVE* activities).

The last action that we need to scrutinize is one that forces qualified operators to select the most appropriate action from among several alternatives (Fig. 6.11).

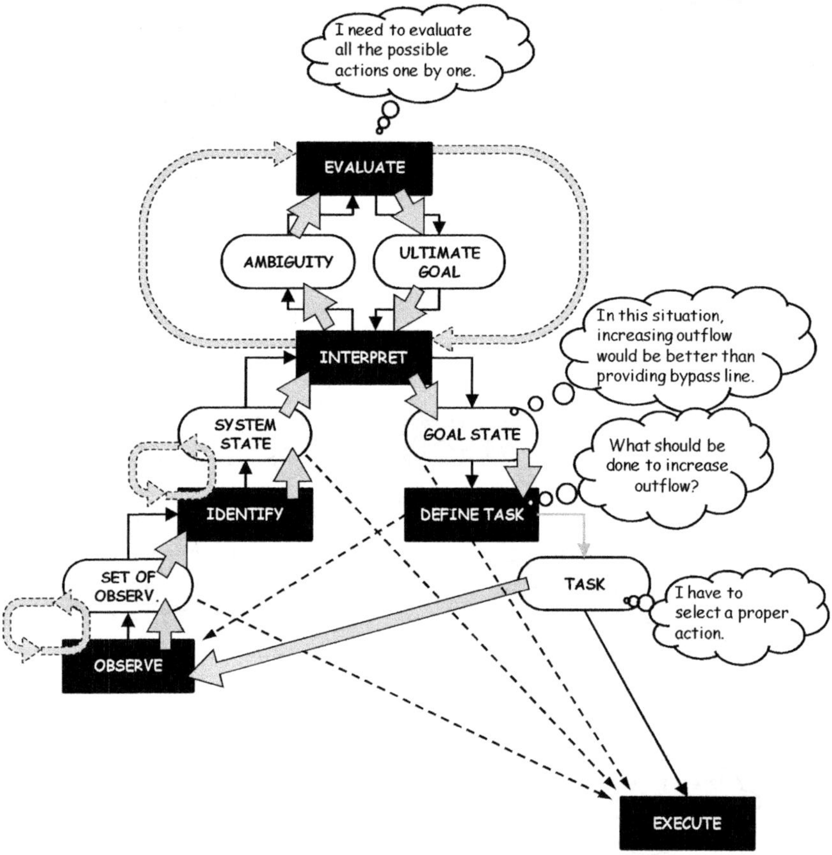

Fig. 6.11 Example of the sequence of decision making activities when qualified operators must select the most appropriate action

Let us recall the fourth action shown in Table 6.11. When qualified operators need to perform this action, they will carry out several activities (i.e., observing necessary information, identifying the state of a related system, etc.) in order to determine whether each alternative is practicable or not in an ongoing situation (e.g., considering the readiness of the associated components or equipment, etc.). Unfortunately, if two or more alternatives are equally probable, then qualified operators should repeatedly evaluate the pros and cons of all possible alternatives. Accordingly, one of the plausible decision making sequences related to the selection of an appropriate action would be illustrated as in Fig. 6.11.

Here, it should be noted that qualified operators have to make one of the three engineering decisions after the selection of an appropriate action. For example, when qualified operators decide that increasing outflow would be better than providing a bypass line, they need to start considering an additional engineering decision to clarify how to increase the outflow. Accordingly, several dotted lines are used in Fig. 6.11 to depict a set of decision making activities related to the performance of the selected action.

In light of the above explanations, we can now characterize engineering decisions. Table 6.12 summarizes the four levels of engineering decisions with the associated meanings. It is to be noted that an action that forces qualified operators to carry out a continuous control is classified as a third level engineering decision (ED-3), because qualified operators need to continuously monitor the satisfaction of an *ACCEPTANCE CRITERION* through the repetition of *IDENTIFY* and *OBSERVE* activities.

Table 6.12 Four levels of engineering decision

Level*	Meaning	Typical action
ED-1	An action that can be accomplished by a simple decision with a clear criterion	Verify the water level of Tank 1 is less than 30%
ED-2	An action that forces qualified operators to integrate lower-level information to create higher-level information	Verify the water level of Tank 1 is decreasing
ED-3	An action that forces qualified operators to identify situations or conditions based on several process parameters, symptoms, and the associated knowledge	Verify the water level of Tank 1 is abnormally decreasing
	An action that forces qualified operators to carry out a continuous control	Maintain the water level of Tank 1 within a range of 23.5% - 50%
ED-4	An action that forces qualified operators to select a proper action	If necessary, perform any of the following. …

*ED: engineering decision.

Based on the above rationales, we can construct an engineering decision graph (EDG) that can be used to characterize the amount of cognitive resources needed to establish the decision criteria of the required actions. For example, Fig. 6.12

depicts EDGs for the first and fourth actions listed in Table 6.11.

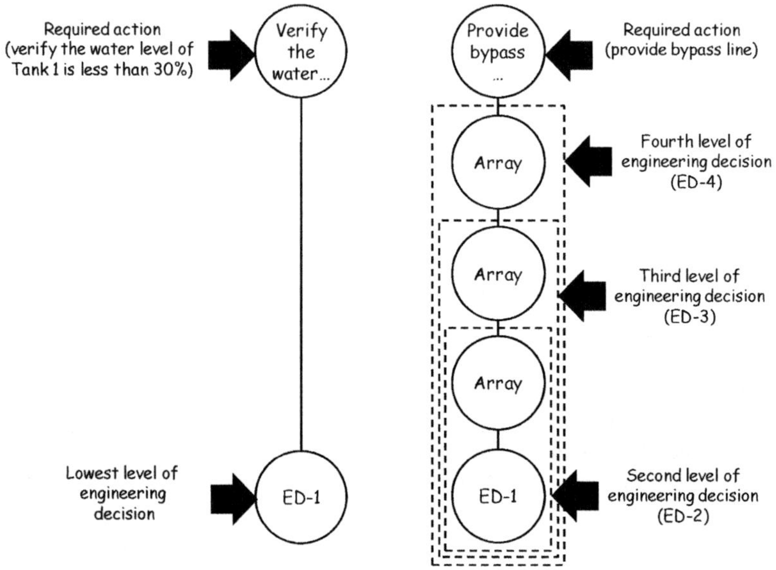

Fig. 6.12 EDGs of two arbitrary actions

As shown in Fig. 6.12, it is assumed that the performance of an action pertaining to ED-2 can be represented by a series of lower level actions (i.e., *array of ED-1*). Similarly, an action related to ED-3 can be expressed by a series of actions belonging to ED-2.

References

Bainbridge L (1997) The change in concepts needed to account for human behavior in complex dynamic tasks. IEEE Trans Syst Man Cybern A Syst Hum 27(3):351–359

Bovair S, Kieras DE (1996) Toward a model of acquiring procedures from text. In: Barr R, Pearson PD, Kamil ML, Mosenthal PB (eds) Handbook of Reading Research, Erlbaum, London, vol 2, pp.206–229

DOD (1999) Human engineering program process and procedures. Mil-HDBK-46855A

DOE (1998) Writer's guide for technical procedures. DOE-STD-1029-92

Dougherty E (1992) SRK – it just keeps on a rollin'. Reliabil Eng Syst Saf 38:253–255

Glossary (2008). Glossary of Terms. http://www.neiu.edu/~dbehrlic/hrd408/glossary.htm

Hollnagel E (1993a) The phenotype of erroneous actions. Int J Man-Mach Stud 39:1–32

Jiang JJ, Klein G (2000) Side effects of decision guidance in decision support systems. Interact Comput 12:469–481

Johnson EJ, Payne JW (1985) Effort and accuracy in choice. Manage Sci 31:395–414

Jung W (2001) Structured information analysis for human reliability assessment of emergency tasks in nuclear power plants. PhD Dissertation, Korea Advanced Institute of Science and Technology, Daejeon, South Korea

Lee JW, Park JC, Lee YH, Oh IS, Lee HC, Jang TI, Kim, Kim DH, Hwang SH, Park JK, Kim JS (2008) Development of the digital reactor safety system. KAERI/RR-2909, Daejeon, South Korea

Moray N (1998) Identifying mental models of complex human-machine systems. Int J Ind Ergonom 22:293–297

Park J, Jung W, Kim J, Ha J (2005) Analysis of human performance observed under simulated emergencies of nuclear power plants. KAERI/TR-2895, Daejeon, South Korea

Rasmussen J (1974) The human data processor as a system component: bits and pieces of a model. Risø-M-1722, Risø Laboratory, Roskilde, Denmark

Rasmussen J (1976) Outlines of a hybrid model of the process plant operator. In: Sheridan TB, Johanssen G (eds) Monitoring Behavior and Supervisory Control, Plenum, New York, pp. 371–383

Rasmussen J (1986) Information processing and human-machine interaction: an approach to cognitive engineering. Elsevier, New York

Shugan SM (1980) The cost of thinking. J Consumer Res 7(2):99–111

Spence JW, Tsai RJ (1997) On human cognition and the design of information systems. Inf Manage 32:73–75

Todd P, Benbasat I (2000) Inducing compensatory information processing through decision aids that facilitate effort reduction: an experimental assessment. J Behav Decis Mak 13:91–106

Ullman DG, D'Ambrosio B (1995) A taxonomy for classifying engineering decision problems and support systems. In: ASME Design Engineering Technical Conferences, Corvalis, OR, vol 2, pp. 627–638

Vicente KJ (1999) Cognitive Work Analysis: Toward Safe, Productive and Healthy Computer-based Work. Erlbaum, Mahwah, NJ

Webster's 2008 New Millennium Dictionary of English Preview Edition (v.0.9.7) http://dictionary.reference.com/browse/action verb

7 Quantifying the Contribution of Task Complexity Factors

In this chapter, practical guidelines to quantifying the contribution of each complexity factor will be explained. In this regard, Table 7.1 that shows an overall quantification scheme will be helpful.

Table 7.1 Eight phases to quantify the contribution of each complexity factor

Phase	Description
1	Extracting the task structure of a procedure
2	Identifying the required actions with the sequence of actions
3	Identifying distinctive actions
4	Identifying necessary information about each distinctive action
5	Assigning the level of domain knowledge to each distinctive action
6	Assigning the level of engineering decision to each distinctive action
7	Constructing four kinds of graphs
8	Quantifying the contribution of each complexity factor

7.1 Extracting a Task Structure

As shown in Table 7.1, the first phase is to identify all the proceduralized tasks as well as the associated procedural steps prescribed in a procedure (i.e., a task structure). In other words, as shown in Fig. 1.1, since a procedure consists of a series of proceduralized tasks containing one or more procedural steps, identifying all the procedural tasks included in the procedure is the first phase in quantifying the complexity of proceduralized tasks. A typical example is Fig. 5.5, which clarifies a part of the task structure of the SGTR procedure of KSNPs (Park et al. 2005).

7.2 Identifying Required Actions with Their Sequence

If the task structure of a procedure being considered is identified, we have to identify the required actions with their sequence. For example, let us look at the following action descriptions that are prescribed in the *Instructions* of the fourth procedural step depicted in Fig. 5.5.

- **IF** pressurizer pressure is less than 123.9 kg/cm^2, **THEN** verify SIAS and CIAS are automatically actuated
- **IF** SIAS and CIAS are **NOT** automatically actuated, **THEN** manually actuate SIAS and CIAS

From the above action descriptions, it seems that the former contains two kinds of required actions, such as *pressurizer pressure is less than 123.9 kg/cm^2* and *verify SIAS and CIAS are automatically actuated.* Similarly, it appears that the latter also consists of two kinds of required actions, such as *SIAS and CIAS are NOT automatically actuated* and *manually actuate SIAS and CIAS.* In addition, since these action descriptions have conditional statements (e.g., a clause followed by IF, THEN, WHEN, WHILE, etc.), it is possible to understand an action sequence to be followed by qualified operators. However, two problems still remain.

The first problem is that some action descriptions do not satisfy the basic requirement of an action description – *each action should consist of one ACTION VERB, an OBJECT, and action specifications.* Figure 7.1 illustrates this problem more clearly.

Original description	Object	Action verb	Action specifications	Remark
Pressurizer pressure is less than 123.9 kg/cm^2	Pressurizer pressure		Less than 123.9 kg/cm^2	• Omitted ACTION VERB
SIAS and CIAS are NOT automatically actuated		SIAS CIAS	NOT automatically actuated	• Omitted ACTION VERB • OBJECT contains two kinds of components having different functions
Manually actuate SIAS and CIAS	SIAS CIAS	Actuate	Manually	• OBJECT contains two kinds of components having different functions
Verify SIAS and CIAS are automatically actuated	SIAS CIAS	Verify	Automatically actuated	

Fig. 7.1 Comparing the basic requirements of an action description

As shown in Fig. 7.1, the description of *verify SIAS and CIAS are automatically actuated* action does not satisfy the basic requirement because it mentions multiple *OBJECT*s, such as SIAS and CIAS, at the same time. In addition, the description of *pressurizer pressure is less than 123.9 kg/cm*^{2} action do not fulfill the basic requirement because there is no *ACTION VERB*.

In order to resolve the problem of having multiple *OBJECT*s in an action description, therefore, we have to subdivide this action into two separate action descriptions that contain a single *OBJECT*. Moreover, the omission of an *ACTION VERB* can be corrected by adopting a hypothetical *ACTION VERB* when the description of an action includes any conditional statement. In other words, since qualified operators have to decide whether a conditional statement is satisfied or not, it is expected that the decision of a conditional statement can be substantiated using appropriate *ACTION VERB*s, such as *determine* or *verify,* for example. Consequently, Table 7.2 summarizes the required actions identified in the fourth procedural step depicted in Fig. 5.5.

Table 7.2 Identifying required actions

Original description	Subdivided action
IF pressurizer pressure is less than 123.9 kg/cm^2, THEN verify SIAS and CIAS are automatically actuated	Determine pressurizer pressure is less than 123.9 kg/cm^2
	Verify SIAS is automatically actuated
	Verify CIAS is automatically actuated
IF SIAS and CIAS are NOT automatically actuated, THEN manually actuate SIAS and CIAS	Determine SIAS is NOT automatically actuated
	Determine CIAS is NOT automatically actuated
	Manually actuate SIAS
	Manually actuate CIAS

The second problem is that some action descriptions seem to be less meaningful because they just represent (or emphasize) the opposite situation of an action. Let us look at two kinds of required actions, such as *verify SIAS is automatically actuated* and *determine SIAS is NOT automatically actuated.* In this case, the latter is unnecessary (or *vice versa*) because the former already encompasses two possible cases – whether SIAS has been automatically actuated or not. In other words, since *verify* forces qualified operators to make a decision, of which the result is either *YES* or *NO*, the whole sequence of required actions can be understood without the latter.

Based on the above explanations, the required actions listed in Table 7.2 can be reduced to the preliminary action sequence shown in Fig. 7.2. It is to be noted that a hypothetical action (i.e., *go to the next procedural step*) is added to Fig. 7.2 because, in most cases, qualified operators have to conduct proceduralized tasks that consist of two or more procedural steps. In addition, the action *perform the fourth procedural step* is added, because each action sequence should have a unique start point.

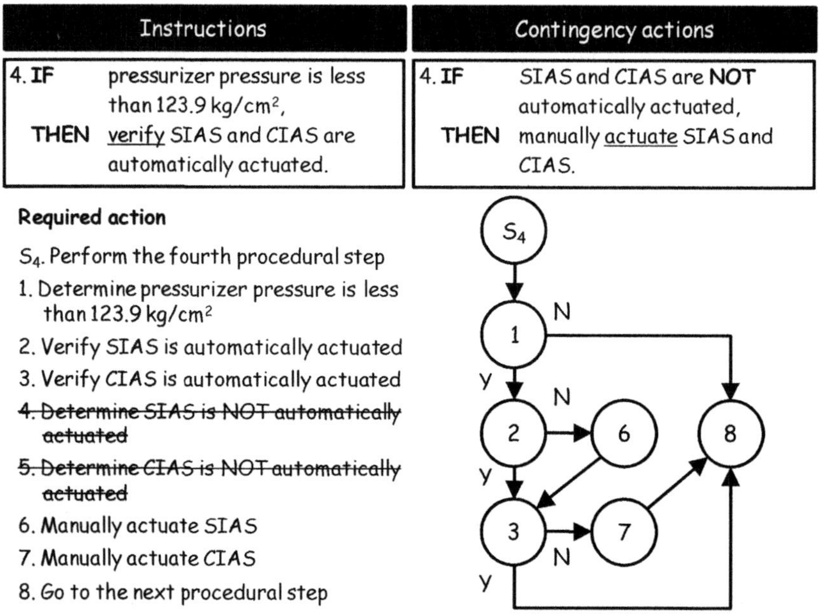

Fig. 7.2 Identifying required actions with their sequence

7.3 Identifying Distinctive Actions

The preliminary sequence of actions presented in Fig. 7.2 is very important for constructing an ACG that can quantify the contribution of two kinds of complexity factors – the number of actions to be conducted by qualified operators and the logical entanglement to be followed by qualified operators. This implies that a set of distinctive actions (DAs) should be carefully identified before constructing an ACG. For example, let us assume a hypothetical ACG with two different procedural steps as depicted in Fig. 7.3.

In Fig. 7.3, each procedural step consists of six actions with the same sequence of actions. This means that the contributions of two kinds of complexity factors about these procedural steps are also identical because they share the same number of actions with the associated sequence. Unfortunately, this result seems to be unrealistic. For example, Kleinsorge et al. (2002) and Mayr and Keele (2000) experimentally showed that the response times of shifting from *Task B* to *Task A* are relatively higher when unqualified operators performed a nonrepeated task set (i.e., *Task C* → *Task B* → *Task A*) instead of a repeated task set (i.e., *Task A* → *Task B* → *Task A*). This strongly supports the notion that the repetition of identical actions will reduce the overall complexity of proceduralized tasks. In this regard, the contribution of complexity factors related to an ACG should be larger when qualified operators conducted procedural step S_1 because there are no repeated actions in it.

For this reason, it is necessary to identify all DAs that are included in a procedure through analyzing the specifications as well as the peculiarity of all the required actions.

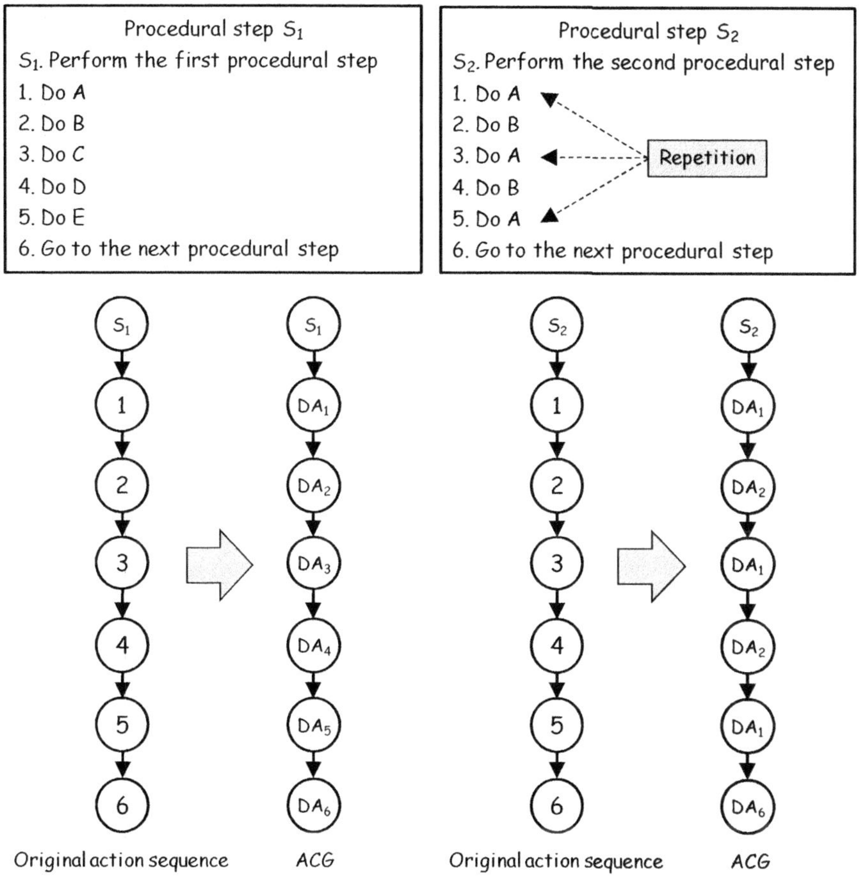

Fig. 7.3 Hypothetical ACGs with two different procedural steps

To this end, Table 7.3 gives an example of a typical usage of an action analysis form, in which distinctive actions can be easily distinguished. For instance, although the original descriptions of two required actions (the second and third actions) are different, they are regarded as the same action (i.e., DA_2), because they share the same action specifications (i.e., the same *OBJECT*, *OBJ*, *INH*, and *NL*) with no *peculiarity*. In contrast, although original descriptions of the first and ninth actions are identical, they should be distinguished as different actions (i.e., DA_1 and DA_8, respectively) because of the *peculiarity* of the ninth action.

Table 7.3 Example of the usage of an action analysis form

	ID	Required action	ACTION VERB	OBJECT	Peculiarity	ACCEPTANCE CRITERION	MEANS	CONSTRAINT
DA_1	1	Stop HPSI pumps	Stop	HPSI pumps	–	OBJ	INH	NL
DA_2	2	Verify pressurizer pressure is less than 123.9 kg/cm^2	Verify	Pressurizer pressure	–	OBJ	INH	NL
	3	Verify pressurizer pressure is between 135.0 and 165.0 kg/cm^2	Verify	Pressurizer pressure	–	OBJ	INH	NL
DA_3	4	Verify pressurizer pressure is abnormally decreasing	Verify	Pressurizer pressure	–	SUB	INH	NL
DA_4	5	Open SBCS valve #1 to 100%, until RCS temperature is less than 260°C	Open	SBCS valve #1	–	OBJ	INH	OBJ_C
DA_5	6	Open SBCS valve #1 to 100%, until RCS temperature is stable	Open	SBCS valve #1	–	OBJ	INH	SUB_C
DA_6	7	Stabilize RCS temperature	Stabilize	RCS temperature	CC	NC	NM	NL
DA_7	8	Depressurize pressurizer pressure using a pressurizer spray valve	Depressurize	Pressurizer pressure	CC	NC	DEG	NL
		IF necessary, perform ANY of the following						
DA_8	9	Stop HPSI pumps.	Stop	HPSI pumps	SEL	OBJ	INH	NL
		…						

*DA is short for distinctive action.

7.4 Identifying Necessary Information

If a set of DAs has been extracted, then the next phase is the identification of necessary information. In other words, all the information to be processed by qualified operators should be identified in this phase. To this end, three kinds of information pertaining to an action specification (i.e., *MEANS*, *ACCEPTANCE CRITERION*, and *CONSTRAINT*) are necessary for performing the required actions. On the basis of these clarifications, Table 7.4 exemplifies the usage of an information analysis form that can identify necessary information.

Table 7.4 Part of an information analysis form

	MEANS[1]	Type[2]	CONSTRAINT	Type	ACCEPTANCE CRITERION	Type
DA_1	HPSI pumps	AAB	–	–	HPSI pumps	AAB
DA_2	Pressurizer pressure	F	–	–	Pressurizer pressure	F
DA_4	SBCS valve #1	AF (jog control)	RCS temperature	F	SBCS valve #1	AF (jog control)

[1]Refer to action descriptions in Table 7.3

[2]Type denotes the basic type of information summarized in Table 6.9

For example, to accomplish DA_1, qualified operators need to manage control-related information (i.e., *MEANS*), which can be determined by the number of HPSI pumps as well as the number of available operating modes. At the same time, qualified operators need the status of HPSI pumps to clarify the *ACCEPTANCE CRITERION* of DA_1. In this regard, since qualified operators are able to directly identify the operating status of HPSI pumps from HPSI pump controllers, *AAB* (*Array of Array of Boolean*) should be commonly regarded as information about *MEANS* as well as about *ACCEPTANCE CRITERION* (Fig. 6.3). Similarly, *AF* (*Array of Float*) is commonly assigned to DA_4, because the source of information about the *MEANS* and the *ACCEPTANCE CRITERION* is a jog controller by which qualified operators are able to not only continuously adjust the open position of the SBCS valve #1 but also identify its open position.

7.5 Assigning the Level of Domain Knowledge

As mentioned in Sect. 3.3.3, qualified operators may feel a cognitive burden if they have to perform an action that requires a high level of domain knowledge. In contrast, qualified operators can probably perform the required action very easily if they are able to accomplish it with a low level of domain knowledge. Accordingly, four levels of domain knowledge are defined in Sect. 6.3.2 based on the Rasmussen's AH framework. Several rules that facilitate the assignment of the levels of domain knowledge are summarized in Table 7.5.

Table 7.5 Several rules for assigning levels of domain knowledge

ID	Rule description
1	The basic level of domain knowledge should be assigned by a knowledge-mapping table.
2	If the objects of the required actions contain the specific property of an entity, then the level of domain knowledge should be determined based on its entity. Typical examples are process parameters or conditions, such as pressurizer pressure, RCS temperature, etc.
3	If the required action does not include any *MEANS* (i.e., *NM*), then the next higher level of domain knowledge compared to the basic level determined from the knowledge-mapping table should be assigned to it.
4	If the *ACCEPTANCE CRITERION* of the required action is *NC* or *SUB*, then the next higher level of domain knowledge compared to the basic level determined from the knowledge-mapping table should be assigned to it.
5	If two or more required actions are grouped by *SEL*, then (1) the next higher level of domain knowledge compared to the basic level of the knowledge-mapping table should be assigned to each action, (2) the highest level of domain knowledge among all the grouped actions should be determined, and (3) the highest level of domain knowledge should be assigned to all the required actions being grouped.
6	AF should be assigned to all the local operations (i.e., *LO*)

The intention of the first rule is to minimize an inconsistency as much as possible, which might be observed during the assignment of the levels of domain knowledge. Table 7.6 shows a typical knowledge-mapping table that could be used for PWRs. For example, if the *OBJECT* of the required action is a kind of pump, then qualified operators should just need domain knowledge pertaining to the function of a component itself (e.g., *CF*). In contrast, if qualified operators have to consider a boundary that consists of two or more components with distinctive functions or purposes, then it is reasonable to anticipate that they will need system level knowledge (e.g., *SF*).

The second rule is related to the assignment of the level of domain knowledge if the *OBJECT* of the required action represents any attribute of it. For example, let us recall DA_2 in Table 7.3, where the *OBJECT* is *pressurizer pressure*. In this case, since pressure is one of the typical attributes of the pressurizer, it is reasonable to assign the level of domain knowledge based on that of the pressurizer. This means that *SF* should be assigned to DA_2 according to the knowledge-mapping table. Similarly, the level of domain knowledge about DA_6 is *PF* because the RCS encompasses several distinctive systems, such as the reactor vessel, RCPs, SGs, etc. In addition, since each system generally has two or more distinctive functions, the concurrent consideration of identical systems should be regarded as *PF*. For example, if the *OBJECT* of an arbitrary action is RCPs (e.g., *stop all RCPs*) or SGs (e.g., *verify all levels of SGs are greater than 23.5%*), we have to assign *PF* to it in order to represent the level of domain knowledge.

Table 7.6 A knowledge-mapping table that could be used for PWRs

Level of domain knowledge	Corresponding object
Component function (CF)	• All kinds of valves, heaters, reservoirs (tanks), batteries, pipes, etc. • All kinds of pumps except RCPs • All kinds of heat exchangers except SGs and condensers • Anything else that can be regarded as a distinguishable functional unit according to a tacit consensus among qualified operators working in PWRs
System function (SF)	• A building such as a containment or turbine building • Reactor vessel • Pressurizer • SGs • RCPs • Diesel generators • Turbines • Condensers • Any boundary that contains two or more distinctive components that have different functions or purposes
Process function (PF)	Any boundary that contains two or more system functions. A typical example is the simultaneous consideration of system functions such as RCPs or SGs
Abstract function (AF)	Any boundary that contains two or more process functions

The third rule implies the enlargement of domain knowledge due to the absence of a proper *MEANS*. Let us look at Fig. 7.4, which compares the changes in an expected problem space of an arbitrary system containing four valves and a reservoir.

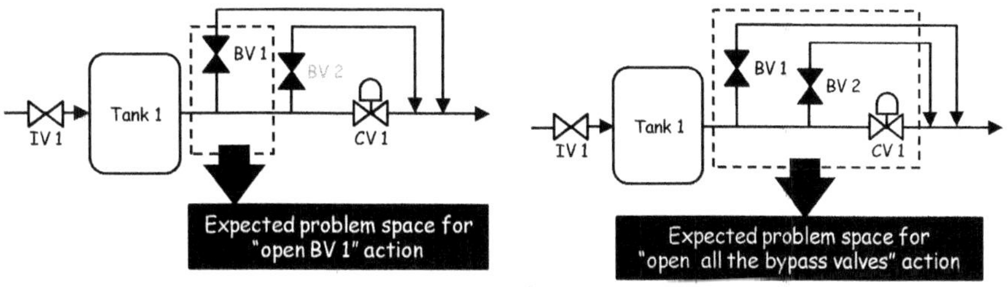

a b

Fig. 7.4 Two examples of the changes in an expected problem space

Above all, as depicted in Fig. 7.4a, it seems to be obvious that qualified operators focus on a narrow problem space to perform *open BV 1* action (refer to an area enclosed by dotted lines) because the *OBJECT* to be acted on is a single component. In contrast, qualified operators probably enlarge their problem space to perform *open all the bypass valves* action because a higher level of domain knowledge will be necessary to answer several questions, such as *which valves are bypass valves*? or *how many bypass valves are linked to Tank 1*?, etc. In other words, as illustrated in Fig. 7.4b, it is anticipated that this action will compel qualified operators to search a certain problem space that consists of several valves surrounding Tank 1. Accordingly, it is reasonable to assume that the next higher level of domain knowledge compared to the basic level determined from the knowledge-mapping table should be assigned to the required action without having detailed specifications about a *MEANS* (i.e., *NM*). This implies that *SF* should be assigned to *open all the bypass valves* action, because the *OBJECT* of this action includes a couple of bypass valves that share the same function (i.e., *CF*).

The fourth rule closely resembles the third rule because the omission of detailed specifications about an *ACCEPTANCE CRITERION* (i.e., *NC*) probably requires additional cognitive resources to process a higher level of domain knowledge. Let us look at Fig. 7.5, which shows a hypothetical trend about the water level of Tank 1.

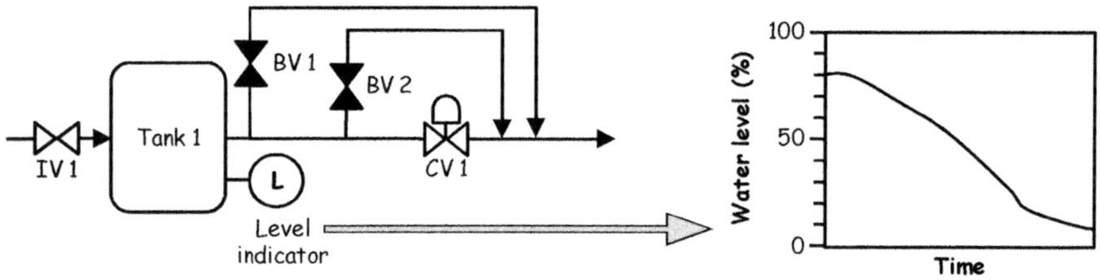

Fig. 7.5 Hypothetical trend in water level of Tank 1

From Fig. 7.5 it is evident that qualified operators can easily perform *verify the water level of Tank 1 is decreasing* action. However, qualified operators are likely to get frustrated when they are faced with *verify the water level of Tank 1 is abnormally decreasing* action because the *ACCEPTANCE CRITERION* of this action varies with respect to the status of surrounding components. That is, if there is no good reason to explain the decrease in the water level of Tank 1, then qualified operators will suspect an abnormal decrease due to other factors, such as a break in a pipe. To this end, qualified operators will carefully observe the status of components that might cause a decrease in the water level of Tank 1, such as the status of BV 1 as well as BV 2 or the position of CV 1 and IV 1, etc. This strongly implies that the fourth rule is meaningful because qualified operators need a higher level of domain knowledge that is indispensable to identifying the associated components to be considered.

The fifth rule is applied when several actions are grouped by *SEL*. For example, let us assume the following equally acceptable actions.

IF necessary, perform **ANY** of the following:
- Stop pump A
- Maintain the water level of pressurizer within 30~50%

In this case, qualified operators have to select the most appropriate action. To do this, as explained in Sect. 6.3.3, qualified operators probably evaluate both actions from many standpoints, such as the suitability of an action for a given situation. From this concern it is natural to assume that qualified operators may need a higher level of domain knowledge compared to an original level assigned by the knowledge-mapping table. Actually, this rule is very similar to both the third and fourth rules because qualified operators need to possess a higher level of domain knowledge to make a decision. However, it is also assumed that the extension of domain knowledge to clarify an effective *MEANS* as well as an ambiguous *ACCEPTANCE CRITERION* (e.g., *SUB* or *NC*) should be different from the selection of the most proper action, because the selection would encompass the evaluation of candidate actions. In other words, since qualified operators have to evaluate not a single action but two or more equally acceptable actions, the total amount of domain knowledge necessary for the selection of the most proper action should be larger than that of a single action with *NM, SUB,* and *NC*. Therefore, the sixth rule is considered in order to compensate for this concern. Fig. 7.6 illustrates detailed steps to explain why the *PF* level is commonly assigned to the above two actions.

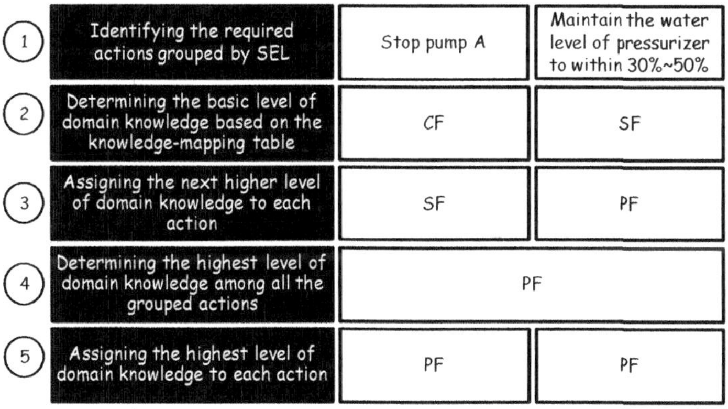

Fig. 7.6 Example illustrating assignment of the levels of domain knowledge when two kinds of required actions are grouped by *SEL*

The last rule concerns actions that require *LO*. As stated in Sect. 6.2.1, it is very difficult to elucidate necessary *MEANS* that would actually be used by field operators. Similarly, it is also difficult to extract an expected problem space to be considered by field operators. However, it seems to be irrational to assign a low level of domain knowledge to this action because higher-level cognitive activities, such as communicating intention between board operators and field operators, are essential for the accomplishment of the required action. Accordingly, for the sake of conservativeness, *AF* is uniformly assumed for actions that require *LO*.

7.6 Assigning the Level of Engineering Decision

After the level of domain knowledge has been assigned, the level of the engineering decision should be assigned. Table 7.7 summarizes several practical rules related to determining the level of the engineering decision.

Table 7.7 Practical rules related to assigning levels of engineering decisions

ID	Rule description
1	The lowest level of the engineering decision (i.e., ED-1) is assigned to an action whose *ACCEPTANCE CRITERION* is *OBJ*, unless its property is not *Trend*
2	ED-1 is assigned to an action whose *CONSTRAINT* is specified by *OBJ_C*
3	The second level of the engineering decision (i.e., ED-2) is assigned to an action if the property of an *ACCEPTANCE CRITERION* is *Trend*
4	The second level of the engineering decision (i.e., ED-2) is assigned to an action if the property of a *CONSTRAINT* is *Trend*
5	ED-2 is assigned to an action whose *ACCEPTANCE CRITERION* is *RI*
6	ED-2 is assigned to an action whose *CONSTRAINT* is *RI_C*
7	The third level of the engineering decision (i.e., ED-3) is assigned to an action if its peculiarity is *CC*
8	ED-3 is assigned to an action whose *ACCEPTANCE CRITERION* is either *SUB* or *NC*
9	ED-3 is assigned to an action whose *CONSTRAINT* is *SUB_C*
10	ED-3 is assigned to an action if there is no specification about *MEANS* (i.e., *NM*)
11	The fourth level of the engineering decision (i.e., ED-4) is assigned to an action if its peculiarity is *SEL*
12	ED-4 is assigned to an action that requires *LO*

For example, let us consider *verify the water level of Tank 1 is less than 30%* action whose *ACCEPTANCE CRITERION* is specified in the form of a discrete value. In this case, qualified operators should be able to easily determine whether the *ACCEPTANCE CRITERION* is satisfied or not. Therefore, this action belongs to the first level of the engineering decision (i.e., ED-1) because a simple decision will be made based on a clear decision criterion.

In addition, the second level of the engineering decision (i.e., ED-2) should be assigned to *verify the water level of Tank 1 is decreasing* action, if we recall that the meaning of ED-2 is an action that forces qualified operators to integrate lower-level information to create higher-level information (Table 6.12). In other words, determining the trend of the water level belongs to ED-2 because qualified operators need to identify the status of the water level by integrating a data series.

Moreover, several rules pertaining to the assignment of the third level (i.e., ED-3) as well as the fourth level of the engineering decision (i.e., ED-4) can be understood in connection with their definitions. For example, let us consider *maintain the water level of Tank 1 within the range 30% to 50% by using CV 1* action in Fig. 7.5. In order to accomplish this action, qualified operators have to answer supplementary questions, such as *how suitable the open position of CV 1 is in this situation*? That is, if the water level is very close to 50%, then qualified operators will be apt to completely close CV 1. In addition, if the change in the water level is not too drastic, then qualified operators will adjust the open position of CV 1 along with the trend of the water level. Obviously, since qualified operators have to establish a proper decision criterion by themselves based on the nature of an ongoing situation, it is meaningful to assign ED-3 to this action. Similarly, if qualified operators have to conduct an action in which there is no specification about *MEANS*, they will probably establish a decision criterion by themselves in order to come up with the proper method for coping with an ongoing situation. Accordingly, it is reasonable to assign ED-3 to this kind of action.

However, the last rule is worthy of special note, because it is assumed valid for the same reason as the assignment of the level of domain knowledge. That is, since it is very difficult to elucidate how field operators can actually perform the required action in a local place, the highest level (i.e., *ED-4*) is assigned for the sake of conservativeness.

7.7 Constructing Four Kinds of Graphs

When all the aforementioned phases are finished, it is possible to construct four kinds of essential graphs through which the contribution of each complexity factor can be quantified by the concept of graph entropies. Let us consider an arbitrary task structure that consists of two procedural steps, $Step_1$ and $Step_2$, as depicted in Fig. 7.7.

		Instructions	Contingency actions
Task (T)	$Step_1$	IF pressurizer pressure is less than 123.9kg/cm^2, THEN verify SIAS and CIAS are automatically actuated.	IF SIAS and CIAS are NOT automatically actuated, THEN manually actuate SIAS and CIAS.
	$Step_2$	IF pressurizer pressure is less than 121.0kg/cm^2 AND SIAS is actuated, THEN perform BOTH of the following: a. Stop ONE RCP in each loop . b. IF RCS subcooling margin is less than 15°C, THEN stop ALL RCPs	

Fig. 7.7 An arbitrary task comprises two procedural steps

First, based on the task structure shown in Fig. 7.7, all the required actions could be identified as listed in Table 7.8. In addition, a set of DAs can be extracted as listed in Table 7.9.

Table 7.8 Required actions included in each procedural step

Procedural step	ID	Required action
$Step_1$	1	Perform $Step_1$
	2	Determine pressurizer pressure is less than 123.9 kg/cm^2
	3	Verify SIAS is automatically actuated
	4	Verify CIAS is automatically actuated
	5	Manually actuate SIAS
	6	Manually actuate CIAS
	7	Go to the next procedural step
$Step_2$	8	Perform $Step_2$
	9	Determine pressurizer pressure less than 121 kg/cm^2
	10	Determine SIAS is actuated
	11	Stop one RCP in each loop
	12	Determine RCS subcooling margin is less than 15°C
	13	Stop all RCPs
	14	Go to the next procedural step

Table 7.9 Action analysis form for the required actions included in $Step_1$ and $Step_2$

DA	ID	ACTION VERB	OBJECT	MEANS	ACCEPTANCE CRITERION	CONSTRA-INT	Pecu-liarity
S_1	1	Perform	$Step_1$	INH	OBJ	NL	–
DA_1	2	Determine	Pressurizer pressure	INH	OBJ	NL	–
	9	Determine	Pressurizer pressure	INH	OBJ	NL	–
DA_2	3	Verify	SIAS	INH	OBJ	NL	–
DA_3	4	Verify	CIAS	INH	OBJ	NL	–
DA_4	5	Actuate	SIAS	INH	OBJ	NL	–
DA_5	6	Actuate	CIAS	INH	OBJ	NL	–
DA_6	7	Go to	Next procedural step	INH	OBJ	NL	–
	14	Go to	Next procedural step	INH	OBJ	NL	–
S_2	8	Perform	$Step_2$	INH	OBJ	NL	–
DA_7	10	Determine	SIAS	INH	OBJ	NL	–
DA_8	11	Stop	(One) RCP	INH	OBJ	RI_C*	–
DA_9	12	Determine	RCS subcooling margin	INH	OBJ	NL	–
DA_{10}	13	Stop	RCPs	INH	OBJ	NL	–

*The specification, such as "in each loop," corresponds to the static configuration (Table 6.5)

Consequently, Fig. 7.8 shows two ACGs for $Step_1$ and $Step_2$ that are constructed based on DAs summarized in Table 7.9.

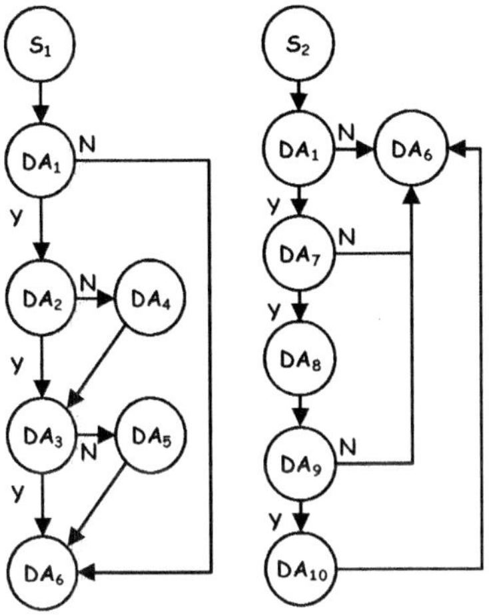

Fig. 7.8 Two ACGs about $Step_1$ and $Step_2$

Second, necessary information to be processed by qualified operators can be identified from DAs. Table 7.10 shows the source of necessary information when qualified operators working in a conventional MCR, have to perform several DAs.

Table 7.10 Information analysis form for $Step_1$ and $Step_2$

ID	MEANS	Type	CONSTRAINT	Type	ACCEPTANCE CRITERION	Type
DA_1	Pressurizer pressure indicator	F	–	–	Pressurizer pressure indicator	F
DA_2 DA_7	SIAS status indicator	B	–	–	SIAS status indicator	B
DA_3	CIAS status indicator	B	–	–	CIAS status indicator	B
DA_4	SIAS actuator	B	–	–	SIAS status indicator	B
DA_5	CIAS actuator	B	–	–	CIAS status indicator	B
DA_8	RCP controller	AB	–	–	RCP controller	AB
DA_9	RCS subcooling margin indicator	F	–	–	RCS subcooling margin indicator	F
DA_{10}	RCP controllers	AAB	–	–	RCP controllers	AAB

Here, there are some points to be noted.

- Necessary information related to S_1, S_2, and DA_6 is not identified because these actions are assumed at our discretion.

- Although the original descriptions of DA_2 and DA_7 are different, the sources of necessary information are the same.
- As SIAS and CIAS can be actuated by a kind of binary controller, their status indicators are necessary to confirm the *ACCEPTANCE CRITERION* (Fig. 6.2a).
- The *CONSTRAINT* of DA_8 is not considered because qualified operators perhaps recall a kind of domain knowledge to perform this action. That is, since information related to identifying *one RCP in each loop* could be extracted from domain knowledge of qualified operators, it is impossible to designate the type of basic information, such as *F* (Float) or *B* (Boolean), etc. Similarly, there are times when it is difficult to identify the types of necessary information if the *ACCEPTANCE CRITERION* or the *CONSTRAINT* of an action has the property such as *equation, formula,* or *dynamic configuration.* Therefore, in order to compensate for this problem, two rules are predefined in Table 7.7. In other words, since the recall of domain knowledge to determine *RI* or *RI_C* could be regarded as the creation of higher level information by integrating lower level information, ED-2 is assigned to an action that contains either *RI* or *RI_C*.

Based on the necessary information summarized in Table 7.10 with the aforementioned notes, we can extract a set of distinctive information (DI) as listed in Table 7.11. This means that qualified operators are supposed to manage at least this kind of information to perform $Step_1$ and $Step_2$. It is to be noted that *RCP controllers* are only considered as DI_6 because the source of information about DA_{10} includes that of DA_8. Accordingly, it is possible to construct two ISGs for $Step_1$ and $Step_2$, as depicted in Fig. 7.9, in which the representation of necessary information will be illustrated by all nodes that are linked to the root nodes, S_1 or S_2.

Table 7.11 Distinctive information identified from $Step_1$ and $Step_2$

		Meaning	Type
$Step_1$	$^*DI_1$	Pressurizer pressure indication	F
	DI_2	SIAS status indication	B
	DI_3	CIAS status indication	B
	DI_4	SIAS actuator	B
	DI_5	CIAS actuator	B
$Step_2$	DI_1	Pressurizer pressure indication	F
	DI_2	SIAS status indication	B
	DI_6	RCP controllers	AAB
	DI_7	RCS subcooling margin indicator	F

*DI: distinctive information

Third, we are able to construct two AHGs for $Step_1$ and $Step_2$ using the list of

DAs and the associated rules to assign the level of domain knowledge. Table 7.12 summarizes the level of domain knowledge assigned to each DA. For example, according to the second rule in Table 7.5, the level of domain knowledge about DA_1 should be SF because pressure is the typical property of a pressurizer.

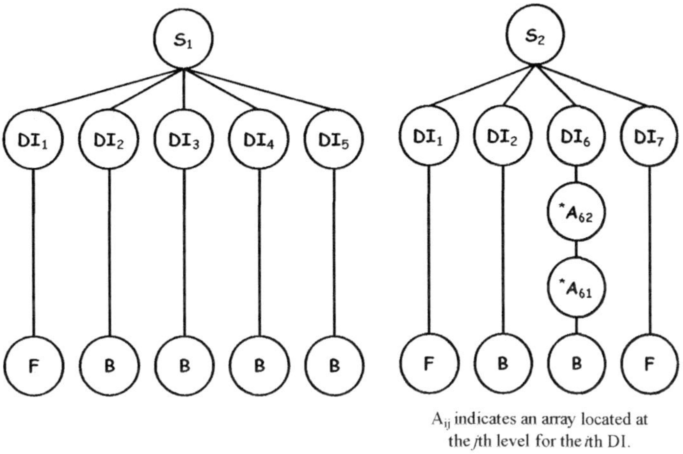

Fig. 7.9 Two ISGs of $Step_1$ and $Step_2$

Table 7.12 Level of domain knowledge of each DA

	DA	Original description	OBJECT	Level of domain knowledge
$Step_1$	DA_1	Determine pressurizer pressure is less than 123.9 kg/cm^2	Pressurizer pressure	SF (pressure is the typical property of a pressurizer)
	DA_2	Verify SIAS is automatically actuated	SIAS	SF (SIAS the typical property of a HPSI system)
	DA_3	Verify CIAS is automatically actuated	CIAS	SF (CIAS is the typical property of a containment)
	DA_4	Manually actuate SIAS	SIAS	SF
	DA_5	Manually actuate CIAS	CIAS	SF
	DA_6	Go to the next procedural step	Next procedural step	CF
$Step_2$	DA_1	Determine pressurizer pressure is less than 121.0kg/cm^2	Pressurizer pressure	SF
	DA_7	Determine SIAS is actuated	SIAS	SF
	DA_8	Stop one RCP in each loop	RCP	SF
	DA_9	Determine RCS subcooling margin is less than 15°C	RCS subcooling margin	PF (subcooling margin is the typical property of a RCS)
	DA_{10}	Stop all RCPs	RCPs	PF
	DA_6	Go to the next procedural step	Next procedural step	CF

Here, it is to be noted that the level of domain knowledge about DA_6 is assumed to be CF. That is, since this action is introduced at our discretion, it is meaningless to consider the level of domain knowledge about DA_6. For this reason, the lowest level of domain knowledge is assigned to DA_6. Figure 7.10 depicts two AHGs for *Step₁* and *Step₂* based on the levels of domain knowledge summarized in Table 7.12.

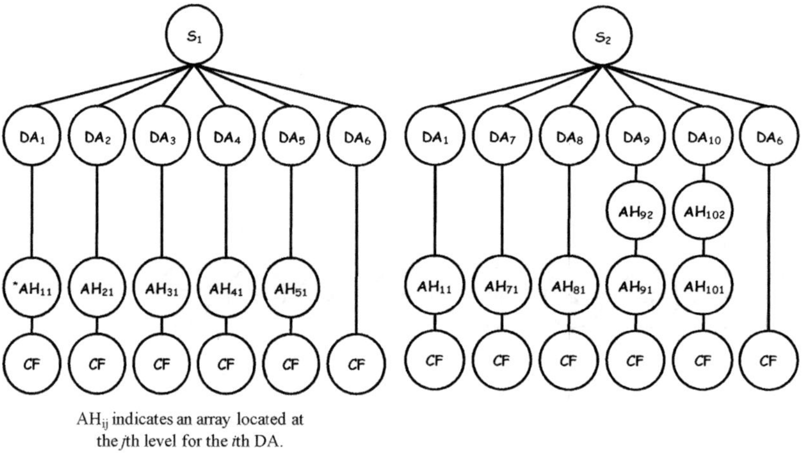

Fig. 7.10 Two AHGs of $Step_1$ and $Step_2$

As for the last graph, two EDGs of $Step_1$ and $Step_2$ can be constructed based on DAs as well as the associated rules to assign the level of the engineering decision. Table 7.13 summarizes the level of engineering decision assigned to each DA.

Table 7.13 Level of engineering decision about each DA

	ID	MEANS	ACCEPTANCE CRITERION	CONSTRAINT	Peculiarity	Assigned level
$Step_1$	DA_1	INH	OBJ	NL	–	ED-1
	DA_2	INH	OBJ	NL	–	ED-1
	DA_3	INH	OBJ	NL	–	ED-1
	DA_4	INH	OBJ	NL	–	ED-1
	DA_5	INH	OBJ	NL	–	ED-1
	DA_6	INH	OBJ	NL	–	ED-1
$Step_2$	DA_1	INH	OBJ	NL	–	ED-1
	DA_7	INH	OBJ	NL	–	ED-1
	DA_8	INH	OBJ	RI_C	–	ED-2
	DA_9	INH	OBJ	NL	–	ED-1
	DA_{10}	INH	OBJ	NL	–	ED-1
	DA_6	INH	OBJ	NL	–	ED-1

For example, according to the first rule given in Table 7.7, the level of engineering decision for DA_1 should be ED-1 because the *ACCEPTANCE CRITERION* of this action is *OBJ*. In addition, the fifth rule in Table 7.7 indicates that the level of the engineering decision for DA_8 should be ED-2 because the *CONSTRAINT* of this action is *RI_C*. In this way, the levels of all the distinctive actions can be systematically determined. As a result, Fig. 7.11 depicts two EDGs for $Step_1$ and $Step_2$.

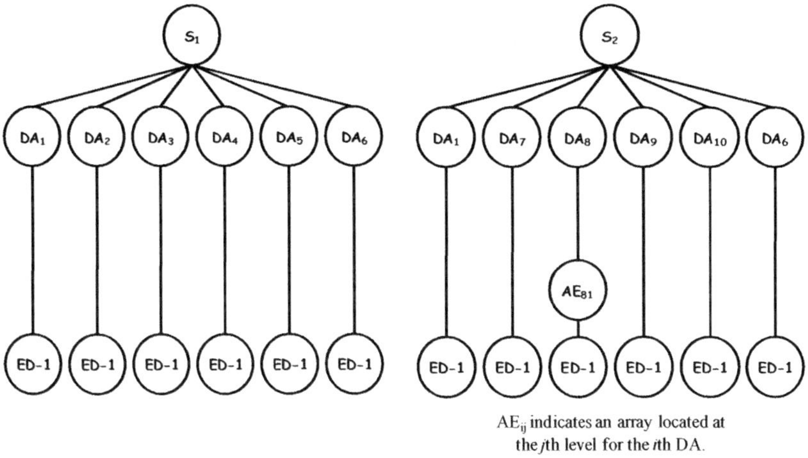

Fig. 7.11 Two EDGs of $Step_1$ and $Step_2$

7.8 Quantifying Five Kinds of Complexity Factors

When all four graphs are constructed, it is possible to quantify the contributions of five kinds of complexity factors based on the associated graph entropies, as clarified in Table 7.14.

Table 7.14 Graph entropies to quantify the associated complexity factors

Complexity factor	Graph entropy
Number of actions	Second-order entropy of an ACG
Logical entanglement	First-order entropy of an ACG
Amount of information	Second-order entropy of an ISG
Amount of domain knowledge	Second-order entropy of an AHG
Level of engineering decision	Second-order entropy of an EDG

For example, let us quantify the contribution of the number of actions in a task depicted in Fig. 7.7. To this end, we need to quantify the second-order entropy of the two ACGs shown in Fig. 7.8. This means that it is essential to introduce the sum of graphs that belong to one of the graph operations.

The sum of two graphs X and Y is mathematically defined as follows (Mowshowitz 1968a): "The sum of X and Y is the graph $X \cup Y$ given by $V(X \cup Y) = V(X) + V(Y)$ and $E(X \cup Y) = E(X) + E(Y)$ where V(X) and E(X) denote the set of vertices (i.e., nodes) and the set of edges (i.e., arcs) included in a graph X, respectively."

Mathematically, the sum of graphs means the simple union of all the nodes as well as the arcs included in all the graphs under consideration. Here, it should be emphasized that there are two rationales supporting the notion that the sum of graphs is meaningful in quantifying the complexity of proceduralized tasks.

First, this concept makes it possible to quantify the contribution of each complexity factor by considering all the necessary graphs of the associated procedural steps without any modification. For example, Fig. 7.12 summarizes the result of node classifications with respect to the sum of two ACGs shown in Fig. 7.8.

The result of node classifications of $Step_1$

Class	Identical node	Neighbor node
I	$\{S_1\}$	$\{DA_1\}$
II	$\{DA_1\}$	$\{S_1, DA_2, DA_6\}$
III	$\{DA_2\}$	$\{DA_1, DA_3, DA_4\}$
IV	$\{DA_3\}$	$\{DA_2, DA_4, DA_5, DA_6\}$
V	$\{DA_4\}$	$\{DA_2, DA_3\}$
VI	$\{DA_5\}$	$\{DA_3, DA_6\}$
VII	$\{DA_6\}$	$\{DA_1, DA_3, DA_5\}$

The result of node classifications of $Step_2$

Class	Identical node	Neighbor node
I	$\{S_2\}$	$\{DA_1\}$
II	$\{DA_1\}$	$\{S_2, DA_6, DA_7\}$
III	$\{DA_6\}$	$\{DA_1, DA_7, DA_9, DA_{10}\}$
IV	$\{DA_7\}$	$\{DA_1, DA_6, DA_8\}$
V	$\{DA_8\}$	$\{DA_7, DA_9\}$
VI	$\{DA_9\}$	$\{DA_6, DA_8, DA_{10}\}$
VII	$\{DA_{10}\}$	$\{DA_6, DA_9\}$

The result of node classifications of the sum of two graphs ($Step_1$ and $Step_2$)

Class	Identical node	Neighbor node
I	$\{S_1, S_2\}$	$\{DA_1\}$
II	$\{^*DA_1\}$	$\{S_1, DA_2, DA_4\}$
III	$\{^*DA_2\}$	$\{DA_1, DA_3\}$
IV	$\{^*DA_3\}$	$\{DA_2, DA_4\}$
V	$\{^*DA_4\}$	$\{DA_1, DA_3\}$
VI	$\{^*DA_5\}$	$\{DA_3, DA_6\}$
VII	$\{^*DA_6\}$	$\{DA_1, DA_3, DA_5\}$
VIII	$\{^{**}DA_1\}$	$\{S_2, DA_6, DA_7\}$
IX	$\{^{**}DA_6\}$	$\{DA_1, DA_7, DA_9, DA_{10}\}$
X	$\{^{**}DA_7\}$	$\{DA_1, DA_6, DA_8\}$
XI	$\{^{**}DA_8\}$	$\{DA_7, DA_9\}$
XII	$\{^{**}DA_9\}$	$\{DA_6, DA_8, DA_{10}\}$
XIII	$\{^{**}DA_{10}\}$	$\{DA_6, DA_9\}$

*A node that belongs to the ACG of $Step_1$.
**A node that belongs to the ACG of $Step_2$.

Fig. 7.12 Distinctive classes to quantify the second-order entropy on the sum of two graphs

As can be seen from Fig. 7.12, two nodes (S_1 and S_2) should be considered identical, because they share the same neighbor node, DA_1. In contrast, it is evident that other nodes do not have the same neighbor node. Accordingly, since the sum of two graphs has a total of 13 distinctive classes, the second-order entropy of ACGs is

$$H_2(Step_1 \cup Step_2) = -\sum_{i=1}^{13} p_i \cdot \log_2 p_i = -\left\{\left(\frac{2}{14}\right)\cdot \log_2\left(\frac{2}{14}\right) + 12\cdot\left(\frac{1}{14}\right)\cdot \log_2\left(\frac{1}{14}\right)\right\} = 3.665.$$

This implies that the contribution of the number of actions on the complexity of a proceduralized task can be quantified as 3.665. In this way, the contributions of other complexity factors on the complexity of proceduralized tasks can be quantified. For the sake of convenience, henceforth, it would be better to define five kinds of submeasures covering the associated complexity factors. These submeasures are given below.

- Step size complexity (SSC), which indicates the complexity due to the number of the required actions to be performed by qualified operators, can be quantified by the second-order entropy of an ACG.
- Step logic complexity (SLC), which denotes the complexity due to the logical entanglement of the required actions, can be quantified by the first-order entropy of an ACG.
- Step information complexity (SIC), which represents the complexity due to the amount of information to be processed by qualified operators, can be quantified by the second-order entropy of an ISG.
- Abstraction hierarchy complexity (AHC), which implies the complexity due to the amount of domain knowledge needed by qualified operators, can be quantified by the second-order entropy of an AHG.
- Engineering decision complexity (EDC), which denotes the complexity due to the amount of cognitive resources for establishing the decision criteria of the required actions, can be quantified by the second-order entropy of an EDG.

Second, the sum of graphs makes it possible to explicitly depict the reduction of a task complexity that stems from the repetition of similar actions. In order to clarify the nature of this characteristic, let us compare the SSC values of three ACGs. In Fig. 7.13, it is observed that two ACGs (i.e., $Step_1 \cup Step_2$) share common graph nodes, DA_1 and DA_6. This means that the value of the SSC about the sum of two ACGs explicitly represents the reduction of a task complexity due to the common graph nodes. According to the theory of graph entropies, the diminution of entropy values due to mutual information (i.e., common graph nodes) is represented by the concept of mutual information (Abramson 1963).

For example, as illustrated in Fig. 7.13, the SSC value about the sum of ACGs is 3.665, while the SSC values of $Step_1$ and $Step_2$ are 2.087 and 2.807, respectively. In theory, the SSC value about the sum of ACGs should be the sum of SSC values of each ACG because the sum of graphs was defined as the simple union of all the nodes as well as the arcs included in all the graphs under consideration. However, since there is mutual information originated from common graph nodes, the actual SSC value of the sum of ACGs is less than the expected value. This implies that the graph entropy value decreases as the number of identical graph nodes increases. Consequently, the complexity of a proceduralized task will decrease in proportion to the number of identical actions to be repeated by qualified operators.

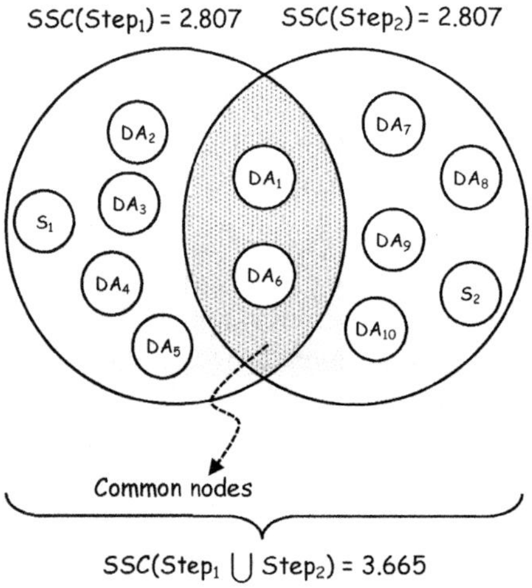

Fig. 7.13 Comparing SSC values of three ACG

References

Abramson N (1963) Information theory and coding. McGraw-Hill, New York

Kleinsorge T, Heuer H, Schmidtke V (2002) Process of task-set reconfiguration: Switching operations and implementation operations. Acta Psychol 111:1–28

Mayr U, Keele SW (2000) Changing internal constraints on action: the role of backward inhibition. J Exp Psychol: Gen 129(1):4–26

Mowshowitz A (1968a) Entropy and the complexity of graphs: I. An index of the relative complexity of a graph. Bull Math Biophys 30:175–204

Park J, Jung W, Kim J, Ha J (2005) Analysis of human performance observed under simulated emergencies of nuclear power plants. KAERI/TR-2895, Daejeon, South Korea

8 Integrating the Contribution of Each Complexity Factor

In Chap. 7, eight phases explaining how to quantify five submeasures were meticulously outlined. Along with these phases, we were able to systematically calculate the contribution of each complexity factor, which is an important clue for evaluating the complexity of proceduralized tasks. However, in order to quantify the complexity of proceduralized tasks, we have to resolve another radical problem – *integrating the five submeasures*. For example, let us consider Table 8.1, which compares the value of each submeasure with respect to arbitrary tasks.

Table 8.1 The values of five submeasures with respect to arbitrary tasks

Sub-measure	Task A	Task B	Task C	Task D	Task E
SSC	3.0	4.0	4.0	4.0	4.0
SLC	4.0	3.0	4.0	4.0	4.0
SIC	4.0	4.0	3.0	4.0	4.0
AHC	4.0	4.0	4.0	3.0	4.0
EDC	4.0	4.0	4.0	4.0	3.0

As highlighted in Table 8.1, the values of the five submeasures have the same composition, such as four identical values with one different value. Here, an interesting question when we look at Table 8.1 would be *are the complexity scores of these tasks equivalent*? In addition, if the complexity scores are not equivalent, a following question would be *how can we properly distinguish them*? This strongly implies that a technical basis should be developed in order to obtain an overall complexity score by integrating all five submeasures. For this reason, it is necessary to introduce a generalized task complexity theory.

8.1 A Generalized Task Complexity Theory

Many researchers have tried to develop a theoretical framework as well as a model representing how to structuralize various kinds of task complexity factors (Campbell 1988; Hackman 1969; Laughlin 1980; McGrath 1984; Roby and Lanzetta

1958; Steiner 1972; Wood 1986; Woods 1988). Of these, the most interesting model would be the one developed by Harvey and his colleagues (Darisipudi 2006; Harvey 2001; Harvey and Koubek 2000; Rothrock et al. 2005). Based on the survey of many existing studies, Harvey and his colleagues suggested a generalized task complexity model that consists of three orthogonal dimensions affecting the complexity of tasks. These dimensions are (1) task scope (TS) representing the breadth, extent, range, or general size of a task being considered, (2) task structurability (TR) indicating whether the sequence as well as the relationship between subtasks are well structured or not, and (3) task uncertainty (TU) pertaining to the degree of predictability or confidence of a task. Based on these definitions, several metrics corresponding to each dimension have been identified, as illustrated in Fig. 8.1.

Dimension	Typical element	Corresponding metric	Remark
TS	Subtasks	Number of subtasks	Subtasks means the decomposed components of a task
	Products	Number of possible products	Products denote the result (or the outcome) of a task
	Product characteristics	Number of ways to measure the success of a product	Any characteristics by which the success of a product can be measured (quality, cost, *etc.*)
	Characteristic conflict	The number of competing product characteristics	Typical examples are the competition between safety and economy or between quality and speed.
	Information	Number of variables	Number of variables to be managed in the course of performing a task
TR	Analyzability	Number of sub-tasks with imperfect mapping to product characteristics	Analyzability would be high if there are clear relations between subtasks and the associated product characteristics
	Alternatives	Number of available paths to reach the desired product characteristics	Multiple paths to reach the desired product characteristics imply a high level of *Alternatives* element
	Coordination	Number of required relations among subtasks	Many kinds of relations among subtasks connote a high level of coordination
TU	Internal confidence	Number of imperfect mappings	The degree of uncertainty or unpredictability due to the structure of subtasks or task alternatives, *etc.*
	External confidence	Number of real- time changes	The level of external confidence would be low if there weremany changes in the required product characteristics
	Random events	Expectation of the number of change occurrences	Random events indicate irregular events accompanying many changes

Fig. 8.1 Three kinds of task complexity dimensions (Park and Jung 2007, © IEEE)

From the point of view of quantifying the complexity of proceduralized tasks, this model is unique because it can be used as a technical basis to integrate the contributions of the five submeasures into a unified measure. In other words, although many researchers have structured various kinds of dominant factors that could make the performance of proceduralized tasks complicated, a model providing the overall structure as well as the dependency among task complexity factors (e.g., the three orthogonal dimensions) seems to be rare. This suggests that it is possible to determine the unified value of a task's complexity by integrating the contribution of each complexity factor. Consequently, as depicted in Fig. 8.2, the unified measure of the complexity of an arbitrary task, called TACOM (TAsk COMplexity), can be regarded as the distance from the origin to an arbitrary point on a one-eighth spherical surface in which TS, TR, and TU have a positive value.

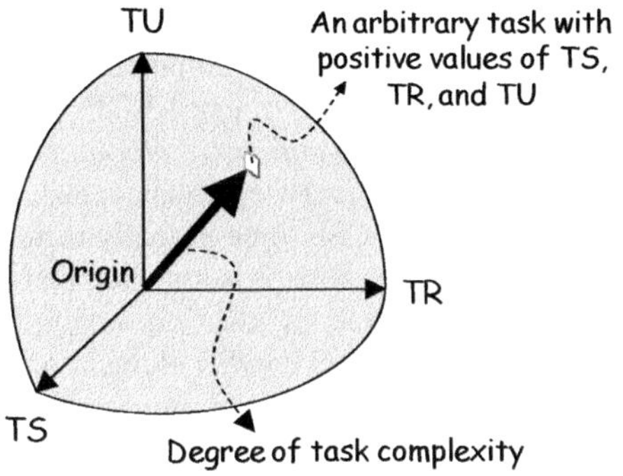

Fig. 8.2 The meaning of the TACOM measure in a hypothetical complexity space created by three orthogonal dimensions

In light of this concern, it is necessary to compare the nature of the five submeasures with the elements considered in the generalized task complexity model. Table 8.2 shows the results of these comparisons.

Table 8.2 Comparing the nature of the five submeasures with typical elements included in the generalized task complexity model (Park and Jung 2007, © IEEE)

Complexity dimension	Typical element	Submeasure
TS	Subtasks	SSC
	Products	–
	Product characteristics	–
	Characteristic conflict	–
	Information	SIC
TR	Analyzability	AHC
	Alternatives	SLC
	Coordination	–
TU	Internal confidence	EDC
	External confidence	–
	Random events	–

8.1.1 TS Dimension

First, it seems to be evident that the SSC that covers the number of the required actions to be done by qualified operators is directly comparable to the *Subtasks* element of the TS dimension. In addition, the SIC pertaining to the amount of information to be processed by qualified operators is congruent with the *Information* element.

In contrast, the other three elements seem to be less meaningful from the point of view of proceduralized tasks, such as emergency tasks. In other words, unlike a dynamic environment in which qualified operators have to accomplish the goal of required tasks without a procedure, two elements (*Product* and *Product characteristics*) should be clarified at the very beginning of an EOP development. For example, one of the ultimate *Products* and *Product characteristics* of EOPs would be *to lead the status of NPPs to a stable condition* and *to minimize radioactive releases into the environment*, respectively. Moreover, every emergency task described in EOPs should have a unique *Product* (e.g., a CSF to be urgently restored) and *Product characteristics* (e.g., allowable time). Therefore, it is assumed that the effects of the two elements on the complexity of emergency tasks are negligible. Similarly, it is assumed that the effect of the *Characteristic conflict* element on the complexity of emergency tasks is also negligible, because the existence of competing *Product characteristics* would be soundly managed in the course of an EOP development.

8.1.2 TR Dimension

Regarding this dimension, it is reasonable to expect that the SLC would be compatible with the *Alternativeness* element because the more the sequence of required actions becomes entangled, the more the number of available paths to accomplish the goal of a given task increases. In addition, the AHC seems to correspond to the *Analyzability* element, because it is anticipated that understanding the cause-and-effect relations between *Subtasks* and their *Product characteristics* would become more difficult in proportion to the amount of domain knowledge needed by qualified operators. To understand this correspondence, let us consider *maintain the water level of Tank 1 lower than 30%* action with two arbitrary systems as depicted in Fig. 8.3.

From Fig. 8.3a, it is not surprising that the *Analyzability* element of the required action is very high. That is, since the only way to control the water level of Tank 1 is to adjust the open position of CV 1, qualified operators can easily confirm the cause (i.e., adjusting CV 1) and the consequence (i.e., the water level of Tank 1). In contrast, the *Analyzability* element of the same action would be low, if qualified operators have to conduct it with the system shown in Fig. 8.3b. That is, it would be not easy to recognize causality without inferring it, because the open positions of three valves would individually or as a group affect the water level of

Tank 1. Therefore, it is reasonable to expect that the *Analyzability* element will decrease along with the increase in the amount of domain knowledge.

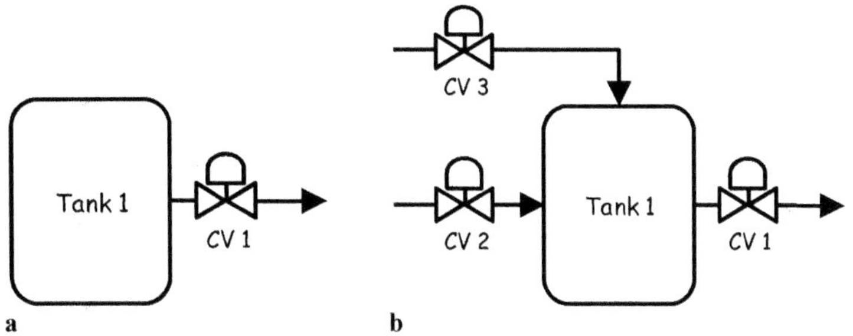

Fig. 8.3 Two arbitrary systems explaining how the amount of domain knowledge affects the *Analyzability* of a given action

However, from the point of view of proceduralized tasks, the consideration of *Coordination* element seems to be unnecessary, because qualified operators are supposed to follow a predefined action sequence to accomplish the goal of a given task. In other words, since the predefined action sequence already contains proper relations among subtasks (i.e., required actions), it is expected that qualified operators will not need to make an effort to organize their sequence.

8.1.3 TU Dimension

In this dimension, it appears that the EDC corresponds to the *Internal confidence* element that is related to the level of uncertainty or unpredictability about required actions. To clarify this aspect, let us recall two actions, DA_1 and DA_8, shown in Table 7.3. Here, it is evident that the level of engineering decision about the former is ED-1 because its *ACCEPTANCE CRITERION* is *OBJ*. In contrast, the level of engineering decision about the latter is ED-4 because qualified operators have to select the most appropriate action based on their own decisions. This strongly suggests that the *Internal uncertainty* element is similar in the nature to the uncertainty related to the level of engineering decision. That is, the degree of uncertainty among task alternatives or subtasks will increase in proportion to the level of engineering decision because qualified operators have to make a decision with a high level of uncertainty, such as *which task alternatives or subtasks should be performed in this situation*?

However, it is assumed that the effect of the *External confidence* element on the complexity of proceduralized tasks is negligible because the real-time changes of *Product characteristics* would be rare when qualified operators perform proceduralized tasks in a procedure. In addition, for the sake of simplicity, it is assumed that the effect of the *Random events* element on the complexity of proceduralized tasks is negligible because it is almost impossible to estimate *how many random events will occur* or *what kinds of random events will occur* in the course of carrying out proceduralized tasks.

8.2 Determining Relative Weights

If we adopt the aforementioned rationales, it is possible to quantify the effect of each complexity dimension on the complexity of an arbitrary task by considering the linear combination of the associated submeasures. For example, the effect of the TS on the complexity of an arbitrary task can be quantified by the linear combination of SIC and SSC with two kinds of relative weights (α_1 and α_2), such as $TS = \alpha_1 \cdot SIC + \alpha_2 \cdot SSC$ $(\alpha_1 + \alpha_2 = 1.0)$. Consequently, we are able to define the TACOM measure with relative weights as depicted in Fig. 8.4.

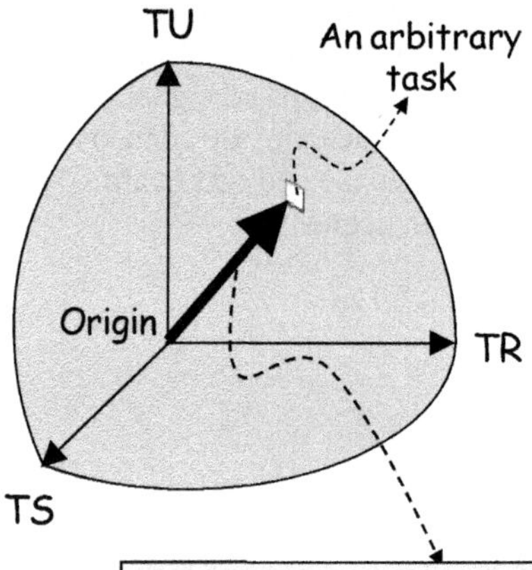

$$TACOM = \sqrt{\alpha \cdot TS^2 + \beta \cdot TR^2 + \gamma \cdot TU^2}$$
$$(\alpha + \beta + \gamma = 1.0)$$
$$TS = \alpha_1 \cdot SIC + \alpha_2 \cdot SSC \ (\alpha_1 + \alpha_2 = 1.0)$$
$$TR = \beta_1 \cdot SLC + \beta_2 \cdot AHC \ (\beta_1 + \beta_2 = 1.0)$$
$$TU = EDC$$

Fig. 8.4 Definition of the TACOM measure

Unfortunately, although there is a technical basis to quantify the complexity of a proceduralized task using the TACOM measure, a crucial problem still remains. That is, it is impossible to obtain TACOM scores without a set of proper weights (i.e., α_1, α_2, β_1, β_2, α, β, and γ). To resolve this problem, we need clarification about the following prerequisites.

8.2.1 Reference Data to for Determining Relative Weights

First, we have to consider what kinds of reference data are meaningful in determining the relative weights of the TACOM measure. As stated in Sect. 2.2, the determination of relative weights could be started from the fact that the increase of the complexity of proceduralized tasks will cause the degradation of the performance of qualified operators. That is, the relative weights of the TACOM measure can be reasonably determined by comparing the performance data of qualified operators.

In this regard, it would be helpful to recall the result of previous studies about the performance measure of a fault diagnosis task (Henneman and Rouse 1984; Henneman and Rouse 1986; Rouse 2007). In order to identify appropriate performance measures related to fault diagnosis tasks, Henneman and Rouse have extensively reviewed various kinds of performance measures including (1) 3 measures addressing the product (results) of diagnosis, (2) 15 measures pertaining to the process of diagnosis, (3) 5 measures of human ability, (4) 3 measures of human aptitude and (5) 4 measures of cognitive styles. They found that all the measures could be grouped into three kinds of basic dimensions: time, error, and inefficiency. Therefore, canonical measures to distinguish a diagnostic performance would be (1) the elapsed time to accomplish a fault diagnosis task, (2) the frequency of incorrect diagnoses, and (3) the number of subdecisions to make a final decision. This means that a prolonged task performance time is a good indication representing the degradation of a diagnostic performance.

Although the characteristics of fault diagnosis tasks are entirely different from those of process control tasks (as well as supervisory control tasks), it is reasonable to assume that the aforementioned performance dimensions could also be valid for representing the performance of qualified operators who have to accomplish proceduralized tasks. Moreover, in the case of carrying out EOPs, time-related data (i.e., task performance time; elapsed time from the commencement of a task to its completion) would be the most meaningful to determine the relative weights of the TACOM measure, for the following two reasons.

First, it is necessary to emphasize that most emergency tasks, especially those prescribed in the early phase of EOPs, were developed based on the well-understood responses of NPPs. At a glance, it might seem to be difficult to evaluate the performance of qualified operators using task performance time, because the ultimate goal is not to carry out emergency tasks as fast as possible but to put NPPs in a stable condition. Accordingly, it may be argued that, *even though qualified operators took a long time to accomplish the required tasks, a prolonged task performance time does not designate the impaired performance of qualified operators*. However, as outlined in Sect. 5.4, several emergency tasks should be accomplished within allowable time limits when the nature of an emergency event is identified (ANS 1994; Chao and Chang 2000; Haas and Bott 1982; Liu et al. 1997; Parzer et al. 1995a; Parzer et al. 1995b, Pearce and Hansen 1986; Roth-Seefrid et al. 1994; Stadelmann and Pappe 1999). For example, one critical emergency task for coping with SGTR events is the isolation of a ruptured SG. The purpose of this task is to stop the increase of the water level in the ruptured SG because the results of previous studies have revealed that the delay of the isolation can trigger a more serious consequence, such as an increased risk of uncontrolled radioactive releases into the environment (Jung et al. 2002; Woods et al. 1990). Therefore, although there is a still uncertainty due to various kinds of determinants (such as leakage rate, break size, physical dimension of the ruptured SG, etc.), it is recommended that the ruptured SG should be isolated within about 30 min. In this case, the delay of the ruptured SG isolation can be regarded as a probe to clarify the impaired performance of qualified operators in the course of performing emergency tasks.

Second, many researchers have experimentally shown that the task performance time is a good measure for elucidating the effect of a cognitive load on the performance of unqualified as well as qualified operators. For example, Fujita (1992) observed that the increase in average task performance time was proportional to the increase in the level of subjective task difficulty. Similarly, Maynard and Hakel (1997) pointed out that time data were sensitive to changes in the level of task complexity measured either objectively or subjectively. In addition, Liu and Wickens (1994) found that task performance time data was useful for evaluating the amount of cognitive demand placed on unqualified operators. Accordingly, if there is a correlation between task performance time and cognitive load, performance time data should be representative of the impaired performance of qualified operators.

8.2.2 Obtaining Task Performance Time Data

If task performance time is meaningful in determining the relative weights of the TACOM measure, then the next concern is very obvious – *how can we obtain task performance time data about emergency tasks*? For this purpose, an operator performance and reliability analysis (OPERA) database developed by the Korea Atomic Energy Research Institute (KAERI) can be used as one of the available data sources (Park and Jung 2005; Park et al. 2005).

The role of the OPERA database is to provide necessary information for scrutinizing human-performance-related problems. To this end, audiovisual records about the retraining sessions of emergency operations have been collected using a full-scope simulator installed in reference NPPs. This full-scope simulator was designed based on the MCR of a 1000 MWe PWR, which consists of conventional control switches, indicators, trend recorders, alarm tiles, etc. In addition, this simulator has been used for the qualifying examination of an operator license, since sufficient verification and validation (V&V) activities have been performed to testify to its functional appropriateness.

It is to be noted that, the retraining course of emergency operations was chosen as the data source of the OPERA database because (1) it is able to secure the performance data of qualified operators during emergencies and (2) it is relatively easy to collect a sufficient number of retraining records, since qualified operators working in the MCR of the reference NPPs must be regularly trained for a period of about 6 months.

As a result of 3 years of data collections, 112 audiovisual records of retraining sessions, which have been conducted by 24 different MCR operating teams, have been gathered, as summarized in Fig. 8.5.

Simulated scenario	Number of simulations	Collection period	Initiating condition
LOCA (loss of coolant accident)	18	Jan. 2000 - Jun. 2000	Initiating conditions were determined by the combination of the following cases; • 11 distinctive break sizes (0.3%, 0.5%, 3%, 4%, 5%, 7%, 10%, 12%, 15%, 20%, 30%) • 9 distinctive break locations
	10	Jan. 2001 - Apr. 2001	
SGTR (steam generator tube rupture)	5	Sep. 1999 - Nov. 1999	
	18	Jul. 2000 - Dec. 2000	
ESDE (excess steam demand event)	18	Jan. 2000 - Jun. 2000	
LOAF (loss of all feed water)	5	Sep. 1999 - Nov. 1999	Loss of all auxiliary feed water pumps (AFWPs)
	18	Jul. 2000 - Dec. 2000	Partial loss of AFWPs
LOOP (loss of off-site power)	10	Jan. 2001 - Apr. 2001	Failure in switchyards
SBO (station black out)	10	Jan. 2001 - Apr. 2001	Failure in diesel generators

Fig. 8.5 Summary of collected records to secure the task performance time data of the reference NPPs (Park and Jung 2007, © Elsevier)

In addition, based on the collected records, a detailed time-line analysis was conducted to extract task performance time data about emergency tasks (Park et al. 2005). Consequently, averaged task performance time data on 91 distinctive emergency tasks were extracted. Appendix B summarizes averaged task performance time data with the associated scores of the five submeasures.

8.3 Determining Relative Weights

As stated earlier, one should be able to determine the relative weights of the TACOM measure based on averaged task performance time data. To this end, one must assume an appropriate fitting model that correlates averaged task performance time with the associated TACOM scores. In light of this concern, the easiest fitting model could be developed based on the assumption of equal weights. However, this assumption seems to be problematic for the following two reasons.

First, existing studies have revealed that the effects of complexity factors on the complexity of tasks are generally not the same. For example, in the case of software complexity, it is well known that the length of source code is the most dominant contributor compared to other complexity factors (Gonzalez 1995; Huang and Lai 1998; Khoshgoftaar et al. 1997; McNicholl and Magel 1982). Accordingly, it is natural to expect that the effects of complexity factors on the complexity of proceduralized tasks would be different.

The second reason is that, in general, a fitting model correlating the performance of unqualified operators with the complexity of a task has a nonlinear form. For example, McNicholl and Magel (1982) stated that "The result of the regression analyses supported our expectation that the Power equation appears to be the best form for capturing the relationship between stimuli and response in our experiment (p. 229)." In addition, Wieringa and Li (1997) mentioned that "The change of presentation of the system may affect human perception of complexity in case the complexity is above a certain threshold (p. 4501)." Actually, this tendency seems to be natural for unqualified operators (or even qualified operators) because it is strongly expected that the amount of available resources would drastically decrease along with the increase of the amount of information to be processed or the increase of the number of actions to be performed (Nowakowska 1986; Salvendy 1997; Wickens 1992).

From the above rationales, therefore, it is possible to determine the relative weights of the TACOM measure by a numerical analysis using a nonlinear fitting model. Subsequently, it is assumed that a fitting model capturing the relationship between task performance time data and the TACOM scores could be explained by an exponential form. As a result, Fig. 8.6 shows a set of relative weights obtained from a nonlinear fitting model with detailed initial conditions as well as constraints. Finally, the TACOM measure with relative weights can be defined as below.

$$\mathrm{TACOM} = \sqrt{0.621\times \mathrm{TS}^2 + 0.239\times \mathrm{TR}^2 + 0.140\times \mathrm{TU}^2}$$

$$\mathrm{TS} = 0.716\times \mathrm{SIC} + 0.284\times \mathrm{SSC}$$

$$\mathrm{TR} = 0.891\times \mathrm{SLC} + 0.109\times \mathrm{AHC}$$

$$\mathrm{TU} = \mathrm{EDC}$$

In addition, Fig. 8.7 depicts the results of a statistical analysis between averaged task performance time data and the associated TACOM scores including the analysis of variance (ANOVA) table.

Fitting model	Time means averaged task performance time data. $Time = a \cdot e^{(b \cdot TACOM)} + c$ $\mathrm{TACOM} = \sqrt{a_3 \times \mathrm{TS}^2 + b_3 \times \mathrm{TR}^2 + c_3 \times \mathrm{TU}^2}$ $\mathrm{TS} = a_1 \times \mathrm{SIC} + a_2 \times \mathrm{SSC}$ $\mathrm{TR} = b_1 \times \mathrm{SLC} + b_2 \times \mathrm{AHC}$ $\mathrm{TU} = \mathrm{EDC}$			
Initial conditions	For fitting model	a = 1.0	b = 1.0	c = 0.0
	For TACOM	$a_3 = 1/3$	$b_3 = 1/3$	$c_3 = 1/3$
	For TS	$a_1 = 0.5$	$a_2 = 0.5$	
	For TR	$b_1 = 0.5$	$b_2 = 0.5$	
Constraints	For fitting model	$a > 0, b > 0, c > 0$		
	For TACOM	$a_3 > 0, b_3 > 0, c_3 > 0$		
		$a_3 + b_3 + c_3 > 0.9999999999$		
		$a_3 + b_3 + c_3 < 1.0000000001$		
	For TS	$a_1 > 0, a_2 > 0$		
		$a_1 + a_2 > 0.9999999999$		
		$a_1 + a_2 < 1.0000000001$		
	For TR	$b_1 > 0, b_2 > 0$		
		$b_1 + b_2 > 0.9999999999$		
		$b_1 + b_2 < 1.0000000001$		
Relative weights	For TACOM	$a_3 = 0.621$	$b_3 = 0.239$	$c_3 = 0.140$
	For TS	$a_1 = 0.716$	$a_2 = 0.284$	
	For TR	$b_1 = 0.891$	$b_2 = 0.109$	

Fig. 8.6 Fitting model, initial conditions and constraints to determine the relative weights of the TACOM measure (Park and Jung 2007, © IEEE)

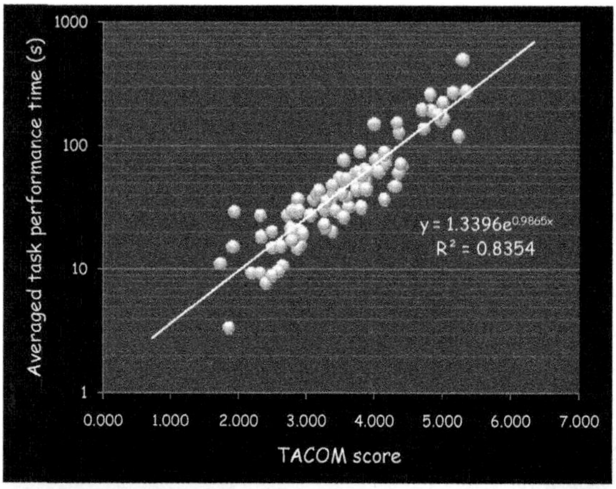

ANOVA table

Item	Degree of freedom	Sum of squares	Mean square	F statistics
Model	1	12.506	12.506	451.969
Error	89	2.463	0.028	
Total	90	14.969		

$F_{0.05}(1, 89) = 3.948$
$p < 10^{-4}$

Residual analysis

- Residual mean: -8.125×10^{-16}
- Normality test: passed ($p = 0.749$)
- Constant variance test: passed ($p = 0.064$)

Fig. 8.7 Result of statistical comparisons between averaged task performance time data and TACOM scores

References

American Nuclear Society (1994) Time response design criteria for safety-related operator actions. ANSI/ANS-58.8-1994

Campbell DJ (1988) Task complexity: A review and analysis. Acad Manage Rev 13(1):40–52

Chao C, Chang C (2000) Development of a dynamic event tree for a pressurized water reactor steam generator tube rupture event. Nuclear Technol 130:27–38

Darisipudi A (2006) Towards a generalized team task complexity model. Ph.D. dissertation, Louisiana State University, Baton Rouge

Fujita Y (1992) Human reliability analysis: a human point of view. Reliabil Eng Syst Saf 38:71–79

Gonzalez RR (1995) A unified metric of software complexity: measuring productivity, quality and value. J Syst Softw 29:17–37

Haas PM, Bott TF (1982) Criteria for safety related nuclear plant operator actions: a preliminary assessment of available data. Reliabil Eng Syst Saf 3:59–72

Hackman JR (1969) Toward understanding the role of tasks in behavioral research. Acta Psychol 31:97–128

Harvey CM (2001) Gauging team tasks: How can one improve the process? In: Proceedings of the Summer Computer Simulation Conference, Orlando, FL, pp.391–396

Harvey CM, Koubek RJ (2000) Cognitive, social and environmental attributes of distributed engineering collaboration: a review and proposed model of collaboration. Hum Factors Ergo-

nom Manuf 10(4):369–393
Henneman RL, Rouse WB (1984) Measures of human problem solving performance in fault diagnosis tasks. IEEE Trans Syst Man Cybern 14(1):99–112
Henneman RL, Rouse WB (1986) On measuring the complexity of monitoring and controlling large-scale systems. IEEE Trans Syst Man Cybern 16(2):193–207
Huang SJ, Lai R (1998) On measuring the complexity of an Estelle specification. J Syst Softw 40:165–181
Jung JH, Chang KS, Kim SJ, Lee JH (2002) Best-estimate analysis of multiple SGTR event in APR 1400 aiming to examine the effect of affected steam generator selection. J Korean Nucl Soc 33(4):358–369
Khoshgoftaar TM, Allen EB, Lanning DL (1997) An information theory-based approach to quantifying the contribution of a software metric. J Syst Softw 36:103–113
Laughlin PR (1980) Social combination processes of cooperative, problem-solving groups as verbal intellective tasks. In: Fishbein M (ed) Progress in Social Psychology, vol 1, Erlbaum, Hillsdale, NJ
Liu TJ, Lin YM, Lee CH, Chang CY, Hong WT (1997) Experimental evaluation of emergency operating procedures on multiple steam generator tube rupture in INER integral system test facility. In: Proceedings of the 8th International Topical Meeting on Nuclear Reactor Thermal-Hydraulics, Kyoto, Japan, pp.1151–1160
Liu Y, Wickens CD (1994) Mental workload and cognitive task automaticity: an evaluation of subjective and time estimation metrics. Ergonomics 37(11):1843–1854
Maynard DC, Hakel MD (1997) Effects of objective and subjective task complexity on performance. Hum Perform 10(4):303–330
McGrath JE (1984) Groups: Interaction and Performance. Prentice Hall, Englewood Cliffs, NJ
McNicholl DG, Magel K (1982) The subjective nature of programming complexity. In: Proceedings of the 1982 Conference on Human Factors in Computing Systems, Gaithersburg, MD, pp.229–234
Nowakowska M (1986) Cognitive Science: Basic Problems, New Perspectives and Implications for Artificial Intelligence. Academic, Orlando, FL
Park J, Jung W (2005) A database for human performance under simulated emergencies of nuclear power plants. Nuclear Eng Technol 37(5):491–502
Park J, Jung W (2007) A study on the revision of the TACOM measure. IEEE Trans Nuclear Sci 5496):2666–2676
Park J, Jung W, Kim J, Ha J (2005) Analysis of human performance observed under simulated emergencies of nuclear power plants. KAERI/TR-2895, Daejeon, South Korea
Parzer I, Petelin S, Mavko B (1995a) Feed-and-bleed procedure mitigating the consequences of a steam generator tube rupture accident. Nucl Engineering Des 154:51–59
Parzer I, Petelin S, Mavko B (1995b) Modelling operator rediagnosis during an SGTR event. Nuclear Eng Des 159:143–151
Pearce RT, Hansen PJ (1986) A generic emergency operations task analysis. In: Proceedings of the International Meeting on Advances in Human Factors in Nuclear Power Systems, Knoxville, TN, pp.336–337
Roby TB, Lanzatta JT (1958) Considerations in the analysis of group tasks. Psychol Bull 55(4):88–101
Rothrock L, Harvey CM, Burns J (2005) A theoretical framework and quantitative architecture to assess team task complexity in dynamic environments. Theor Issues Ergonom Sci 6(2):151–171
Roth-Seefrid M, Feigel A, Moser HJ (1994) Implementation of bleed and feed procedures in Siemens PWRs. Nuclear Eng Des 148:133–150
Rouse WB (2007) People and Organization: Explorations of Human Centered Design. Wiley, Hoboken, NJ
Salvendy G (1997) Handbook of Human Factors and Ergonomics, 2nd edn. Wiley, New York
Stadelmann W, Pappe W (1999) State-oriented accident management and emergency procedures

at Gundremmingen nuclear power plant. Kerntechnik 64(3):107–117
Steiner ID (1972) Group Process and Productivity. Academic, New York
Wickens CD (1992) Engineering psychology and human performance, 2nd edn. Harper Collins, New York
Wieringa PA, Li K (1997) Reducing operator perceived complexity. In: Proceedings on IEEE International Conference on Computational Cybernetics and Simulation, Orlando, FL, vol 5, pp.4498–4502
Wood RE (1986) Task complexity: definition of the construct. Organizat Behav Hum Decis Processes 37:60–82
Woods DD (1988) Coping with complexity: the psychology of human behavior in complex systems. In: Goodstein LP, Anderson HB, Olsen SE (eds) Tasks, Errors and Mental Models, Taylor and Francis, London, pp.128–148
Woods DD, Roth EM, Pople, HE Jr. (1990) Modeling operator performance in emergencies. In: Proceedings on the 34th Human Factors and Ergonomics Society Annual Meeting, Orlando, FL, pp.1132–1136

9 Validation of TACOM Measure

From the previous chapter, the TACOM measure is now available to quantify the complexity of proceduralized tasks. Therefore, the last question about the development of the TACOM measure would be: *is the TACOM measure meaningful for quantifying the complexity of proceduralized tasks*?

In order to answer this question, we can consider two kinds of validation. The first one is to directly compare the performance of qualified operators with the associated TACOM scores. That is, one should be able to validate the appropriateness of the TACOM measure from the point of view of three performance dimensions – *time, error,* and *efficiency*. The second kind of validation can be deduced from one of the canonical advantages of a good procedure. As stated in Sect. 2.1, good procedures guarantee at least three major advantages, and one of them is the standardization of the performance of qualified operators. This means that if the TACOM measure can quantify the complexity of proceduralized tasks, then the performance of qualified operators should be similar when they are performing proceduralized tasks with similar TACOM scores.

9.1 Validation Activity – Outline

Let us look at Fig. 9.1, which illustrates the overall validation scheme regarding the appropriateness of the TACOM measure. In Fig. 9.1, detailed activities belonging to the first validation aspect correspond to TACOM scores vs. three kinds of performance data that represent the basic performance dimensions. Unfortunately, since the error rate of qualified operators is generally low, it is very difficult to collect a sufficient amount of error-related data. In addition, since the relative weights that are indispensable for quantifying TACOM scores have been determined by averaged task performance time data, it is reasonable to expect that there would be a significant correlation between averaged task performance time data and TACOM scores. For this reason, the only viable activity would be comparing TACOM scores with subjective workload scores to reflect the *inefficient* dimension.

Meanwhile, the validation activities belonging to the second category are very straightforward because the standardization aspect of the TACOM measure will be clarified by comparing TACOM scores with the associated performance data that were gathered not only from the reference NPPs but also from other NPPs. Unfortunately, although the standardization aspect should be clarified from the other two dimensions (i.e., the *error* and the *inefficiency*), the only viable activity seems to be comparing averaged task performance time data (i.e., the *time* dimension) due to the difficulty in securing the associated performance data.

Validation aspect	Hypothesis		Validation activity	Remark
Task performance	Task performance will decrease with respect to the increase of the complexity of proceduralized tasks (i.e., the TACOM score)	Time	Comparing TACOM scores with averaged task performance time data	Already compared in order to determine relative weights
		Error	Comparing TACOM scores with error rates	It is difficult to secure sufficient amount of data
		Inefficiency	Comparing TACOM scores with subjective workload scores	Viable activity
Standardization	Task performance will remain in a certain range if qualified operators carry out proceduralized tasks that have similar TACOM scores	Time	Comparing TACOM scores with averaged task performance time data gathered from different NPPs	
		Error	Comparing TACOM scores with error rates gathered from different NPPs	It is difficult to secure sufficient amount of data
		Inefficiency	Comparing TACOM scores with subjective workload scores gathered from different NPPs	

Fig. 9.1 Validation scheme of TACOM measure

9.2 Comparing with Subjective Workload Scores

9.2.1 NATA–TLX Technique

As stated by Henneman and Rouse (1984), the diagnostic performance of qualified operators will be ineffective if they reach a final decision through many subdecisions. This means that qualified operators who follow an ineffective way of thinking are likely to feel a high level of cognitive demand compared to those who follow an effective way of thinking, because the former expended more efforts than the latter. Thus, it is necessary to emphasize that a subjective workload is susceptible to a certain level of cognitive demands (Campbell 1988). This strongly suggests that a subjective workload would be a good indicator to represent the *inefficiency* dimension of human performance. In addition, since the amount of effort to be spent by qualified operators will increase as task complexity increases, the subjective workload should increase in proportion to the complexity of tasks to be performed (Stassen et al. 1990; Maynard and Hakel 1997; Li and Wieringa 2000; Hancock 1996; Wei et al. 1998).

Therefore, although many researchers have criticized the meaning of subjective workload scores, the TACOM measure can be regarded as a proper indicator of the complexity of proceduralized tasks, if there is a tendency whereby subjective workload scores increase as TACOM scores increase. For this reason, TACOM scores and subjective workload scores are compared in order to investigate the appropriateness of the TACOM measure from the point of view of the *inefficient* dimension.

Many kinds of subjective workload measurement techniques have been developed in recent decades (Vidulich and Tsang 1986; Nygren 1991; Dickinson et al. 1993; Hendy et al. 1993; Hancock 1996; Svensson et al. 1997; Hill et al. 1992). Of these, the NASA–TLX (National Aeronautics and Space Administration – task load index) technique has been selected as the reference method to measure subjective workload scores because it (1) provides detailed as well as diagnostic results (Hill et al. 1992), (2) is able to support the general prediction model for a subjective workload (Nygren 1991), and (3) is known as one of the most suitable techniques for evaluating the level of subjective workloads (Liu and Wickens 1994).

The NASA–TLX technique was first developed in the 1980s (Hart and Staveland 1988), and it quantifies a subjective workload by a weighted average of ratings on six dimensions, such as mental demand (MD), physical demand (PD), temporal demand (TD), performance (PE), effort (EF), and frustration (FR) (NASA 2009). To this end, the evaluators are asked to identify the relative weights of six dimensions about the workload of a given task based on their knowledge and experience. Then, the evaluators are asked to assess subjective scores about six dimensions using an arbitrary scale ranging from 0 to 100, which represent the level of subjective workload they felt in the course of performing the required task. Finally, based on the relative weights and subjective ratings, the overall workload can be quantified by their weighted average:

$$\mathrm{NASA-TLX} = a_1 \times \mathrm{MD} + a_2 \times \mathrm{PD} + a_3 \times \mathrm{TD} + a_4 \times \mathrm{PE} + a_5 \times \mathrm{EF} + a_6 \times \mathrm{FR}$$

where a_i ($i = 1, \ldots, 6$) denotes the relative weight

However, since evaluators have to follow a quite tricky process to determine relative weights (Hart and Staveland 1988), an equally weighted average has been suggested as an alternative method, such as $a_i = 1/6$ (Nygren 1991).

9.2.2 Gathering Subjective Workload Scores

In order to gather subjective workload scores pertaining to the performance of emergency tasks, SROs working in the MCR of the reference NPPs were chosen, for two reasons. First, it is reasonable to assume that most of the burden that may arise in the course of performing emergency tasks will be loaded on the SRO of each operating team, because the SRO is responsible for the performance of emergency tasks (Moray 1999; Reinartz and Reinartz 1992). As outlined in Sect. 5.5, most of the actions included in emergency tasks should be carried out by the command as well as the confirmation of SROs. Under this operation scheme, it seems to be less meaningful to consider the subjective workload of board operators (i.e., ROs, TOs, and EOs).

Second, it should be emphasized that SROs have sufficient experience with emergency tasks prescribed in EOPs owing to regular retraining (for a period of about 6 months) for various kinds of initiating conditions. In other words, since the NASA–TLX technique quantifies a subjective workload based on personal experience with a given task to be evaluated, it is essential to select qualified operators who are familiar with the performance of emergency tasks. From these concerns, in total 18 SROs were asked to rate 6 dimensions about 23 emergency tasks that had been selected from the EOPs of reference NPPs. Table 9.1 summarizes the list of selected emergency tasks.

Table 9.1 Emergency tasks selected from the reference NPPs (Park and Jung 2006, © IEEE)

ID	Corresponding EOP	Procedural step		Remark
		Start	End	
1	ESDE (excess steam demand event)	4.0	5.0	–
2	LOCA (loss of coolant accident)	6.0	7.0	Group A
3	ESDE	7.0	8.0	Group A
4	ESDE	13.0	16.0	Group B
5	ESDE	17.0	18.0	–
6	SGTR	6.0	7.0	Group A
7	ESDE	24.0	28.0	–
8	ESDE	29.0	30.0	–
9	SGTR	8.0	10.0	–
10	SGTR	11.0	14.0	–
11	LOCA	11.0	13.0	–
12	LOCA	21.0	24.0	Group B
13	LOCA	15.0	19.0	–
14	ESDE	37.0	38.0	Group C
15	LOOP (loss of off-site power)	3.0	4.0	–
16	SGTR	15.0	18.0	Group B
17	LOOP	8.0	13.0	–
18	LOCA	27.0	28.0	Group C
19	LOAF (loss of all feed water)	5.0	10.0	–
20	LOAF	11.0	16.0	–
21	SBO (station blackout)	4.0	6.0	–
22	SBO	7.0	13.0	–
23	SBO	14.0	18.0	–

In Table 9.1, *Start* and *End* in the *Procedural step* column refer to procedural steps that denote, respectively, the commencement and the accomplishment of a given emergency task. For example, the first task is started from the fourth procedural step of the ESDE procedure, and then completed when the performance of the fifth procedural step has been finished. It is to be noted that the meaning of the three groups in the *Remark* column of Table 9.1 will be explained later.

On the basis of the selected emergency tasks, eight tasks were assigned to each SRO by the following sequence: (1) three emergency tasks belonging to *Groups A, B, and C* were evenly assigned and (2) the remaining emergency tasks not belonging to the three groups were randomly assigned. Table 9.2 summarizes the emergency tasks assigned to each SRO.

Table 9.2 Emergency tasks assigned to each SRO (Park and Jung 2006, © IEEE)

SRO ID	Task ID about 8 tasks assigned to each SRO							
1	3	4	9	11	13	14	17	23
2	1	3	4	5	8	18	20	23
3	1	3	9	12	14	19	22	23
4	3	7	9	12	15	18	19	22
5	3	5	8	14	15	16	17	20
6	1	3	9	15	16	18	20	23
7	2	4	8	11	13	14	15	21
8	2	4	7	10	11	13	18	23
9	2	5	7	10	12	14	15	19
10	2	8	9	10	12	13	18	22
11	1	2	5	11	14	16	17	21
12	2	5	10	16	18	19	21	23
13	4	6	7	10	13	14	17	20
14	1	4	5	6	8	18	20	21
15	6	10	12	14	17	19	21	22
16	1	6	7	9	12	17	18	22
17	6	8	11	14	15	16	19	22
18	6	7	11	13	16	18	20	21

Then, SROs gave subjective scores on six dimensions, which represent the amplitude of the workload they felt in the course of performing the assigned emergency tasks. Consequently, Table 9.3 shows subjective workload scores with the associated emergency tasks. It is to be noted subjective workload scores appearing in the each row of Table 9.3 indicate all the NASA–TLX scores given by SROs who were asked to assess emergency tasks. Accordingly, since a total of nine SROs participated in the evaluation of the 14th and 18th emergency tasks (refer to *Group C* in Table 9.1), those tasks have two more NASA–TLX scores than the others. In addition, *Average* represents the mean value of NASA–TLX scores for a given emergency task.

Table 9.3 Summary of subjective workload scores (Park and Jung 2006, © IEEE)

Task ID	Average	Subjective workload score								
1	38.1	34.2	69.2	29.2	40.0	35.0	20.8	–	–	–
2	41.3	51.7	46.7	38.3	43.3	29.2	38.3	–	–	–
3	44.7	55.0	31.7	58.3	43.3	51.7	28.3	–	–	–
4	45.6	48.3	35.0	55.0	50.0	40.0	45.0	–	–	–
5	46.3	41.7	56.7	47.5	43.3	35.0	53.3	–	–	–
6	38.8	40.0	41.7	44.2	30.0	43.3	33.3	–	–	–
7	53.9	49.2	62.5	63.3	48.3	55.0	45.0	–	–	–
8	52.2	60.0	35.0	55.0	65.8	38.3	59.2	–	–	–
9	55.0	65.0	71.7	53.3	30.0	48.3	61.7	–	–	–
10	54.6	63.3	54.2	50.0	41.7	61.7	56.7	–	–	–
11	52.9	45.0	37.5	63.3	55.0	55.0	61.7	–	–	–
12	43.1	60.0	38.3	38.3	41.7	42.5	37.5	–	–	–
13	48.6	44.2	51.7	60.8	43.3	51.7	40.0	–	–	–
14	53.9	58.3	69.2	26.7	65.0	43.3	61.7	56.7	45.8	58.3
15	47.9	61.7	24.2	60.0	30.8	56.7	54.2	–	–	–
16	39.5	48.3	24.2	36.7	35.0	58.3	34.2	–	–	–
17	47.1	45.0	51.7	55.0	43.3	27.5	60.0	–	–	–
18	48.8	36.7	28.3	55.0	65.0	45.0	46.7	62.5	58.3	41.7
19	55.7	61.7	67.5	40.0	57.5	40.0	67.5	–	–	–
20	49.4	45.8	46.7	58.3	30.8	55.0	60.0	–	–	–
21	63.7	35.8	65.0	73.3	55.0	82.5	70.8	–	–	–
22	61.3	65.0	79.2	58.3	51.7	70.0	43.3	–	–	–
23	51.0	56.7	42.5	66.7	38.3	51.7	50.0	–	–	–

9.2.3 Reliability of Subjective Workload Scores

As summarized in Table 9.3, NASA–TLX scores on 23 emergency tasks have been successfully obtained. However, before comparing NASA–TLX scores with the associated TACOM scores, it is essential to check their reliability. In this regard, it is necessary to consider two aspects related to the reliability of subjective ratings – *consistency* and *reproducibility*.

First, the consistency (or the agreement) of NASA–TLX scores should be clarified because SROs' ratings on six dimensions could be changed for various reasons, such as aptitude or personality, for example. In other words, if SROs' ratings fluctuate due to factors besides the performance of emergency tasks, the reliability of NASA–TLX scores would be questionable. From this concern, an intraclass correlation (ICC) coefficient was used to confirm the consistency of SROs' ratings (Bartko 1966; Bartko 1976).

The ICC coefficient ranges from $-\infty$ to 1, and the level of consistency increases with increases in the ICC coefficient. Accordingly, one indicates perfect consistency, while a negative value of the ICC coefficient denotes that subjective ratings are unreliable because of the lack of consistency. Table 9.4 summarizes the classes of ICC coefficients that have been frequently adopted as a basis for determining the consistency level of subjective ratings (Landis and Koch 1977).

Table 9.4 Levels of consistency of subjective ratings

Level of consistency	Corresponding ICC coefficient
Poor	Negative value
Slight	0 to 0.2
Fair	0.21 to 0.4
Moderate	0.41 to 0.6
Substantial	0.61 to 0.8
Almost perfect	0.81 to 1.0

In addition, the result of existing studies found that subjective ratings would be consistent when their ICC coefficient locates at least in the moderate level (Landis and Koch 1977; Marinus et al. 2004). Consequently, 0.41 is used as the threshold value from which the consistency of NASA–TLX scores can be determined. As a result, Table 9.5 summarizes TACOM scores as well as the associated NASA–TLX scores with the ICC coefficients of all the emergency tasks. It is to be noted that a strikethrough in Table 9.5 indicates an emergency task having an unreliable NASA–TLX score.

Second, the reproducibility (or repeatability) of NASA–TLX scores should be considered in order to confirm the reliability of subjective ratings (Bruton et al. 2000; Levy et al. 1999). In other words, even if there is consistency, if SROs assigned different scores to the same emergency tasks, then it may be difficult to use the collected NASA–TLX scores as the reference data to validate the appropriateness of the TACOM measure. Therefore, in order to clarify the reproducibility, it is necessary to internally compare NASA–TLX scores of the same emergency tasks. To this end, three groups of emergency tasks are selected and then randomly assigned to SROs, as noted in Table 9.2 (i.e., *Groups A, B, and C*).

For example, let us consider the second, third, and sixth emergency tasks in Table 9.1, which belong to *Group A*. Here, the goal of the sixth emergency task is *checking the necessity of stopping RCPs*, which consists of two procedural steps prescribed in the SGTR procedure, as illustrated in Fig. 5.5. The interesting point is that, in order to accomplish the same goal, identical procedural steps are also stipulated in both a LOCA (i.e., the second emergency task) and an ESDE procedure (i.e., the third emergency task).

Table 9.5 TACOM scores, NASA–TLX scores, and ICC coefficients

Task ID	TS	TR	TU	TACOM	Average	ICC
~~1~~	~~4.688~~	~~2.506~~	~~5.012~~	~~4.321~~	~~38.10~~	~~0.33~~
2	4.868	2.160	3.784	4.223	41.25	0.77
3	4.868	2.160	3.784	4.223	44.73	0.41
4	4.841	2.526	5.223	4.461	45.57	0.50
5	4.586	1.765	6.393	4.419	46.30	0.51
6	4.868	2.160	3.784	4.223	38.73	0.49
7	5.973	2.757	6.624	5.488	53.90	0.48
8	5.481	2.471	5.306	4.905	52.20	0.41
~~9~~	~~5.711~~	~~2.792~~	~~6.515~~	~~5.297~~	~~55.00~~	~~0.37~~
10	6.089	2.407	6.355	5.483	54.58	0.53
~~11~~	~~5.293~~	~~2.708~~	~~4.884~~	~~4.742~~	~~52.92~~	~~0.39~~
12	4.841	2.526	5.223	4.461	39.43	0.53
13	5.502	2.494	6.442	5.108	48.61	0.47
14	5.881	2.235	6.731	5.386	53.85	0.44
~~15~~	~~5.387~~	~~2.645~~	~~3.889~~	~~4.670~~	~~47.92~~	~~0.33~~
16	4.841	2.526	5.223	4.461	43.08	0.42
17	5.717	2.403	7.083	5.357	47.08	0.46
18	5.881	2.235	6.731	5.386	48.78	0.43
~~19~~	~~5.871~~	~~2.854~~	~~6.204~~	~~5.361~~	~~55.69~~	~~0.38~~
~~20~~	~~6.064~~	~~2.392~~	~~7.026~~	~~5.578~~	~~49.44~~	~~0.38~~
~~21~~	~~4.768~~	~~2.021~~	~~3.866~~	~~4.145~~	~~63.75~~	~~0.38~~
22	5.727	2.675	6.091	5.222	61.25	0.46
23	5.120	2.473	5.266	4.650	50.97	0.42

This means that the reproducibility can be investigated by comparing whether or not SROs give similar NASA–TLX scores to the same emergency tasks. Based on this concern, Table 9.6 shows the results of one-way ANOVA conducted for three groups of emergency tasks.

From Table 9.6 it seems to be evident that there is no significant difference among NASA–TLX scores for the three groups of emergency tasks. For example, the mean values of NASA–TLX scores for the three kinds of emergency tasks belonging to *Group A* are similar because their ANOVA result strongly indicates that the difference among NASA–TLX scores is due to random variability (i.e., $p = 0.54$). Similarly, the ANOVA results of other groups indicate that SROs have given similar NASA–TLX scores when they are asked to rate the same emergency tasks. Consequently, one could reasonably expect reproducibility of NASA–TLX scores.

Table 9.6 ANOVA results of three groups of emergency tasks (Park and Jung 2006, © IEEE)

Group	Task ID	Corresponding NASA –TLX score rated by SROs									p^*
A	2	51.7	46.7	38.3	43.3	29.2	38.3	–	–	–	0.54
	3	55.0	31.7	58.3	43.3	51.7	28.3	–	–	–	
	6	40.0	41.7	44.2	30.0	43.3	33.3	–	–	–	
B	4	48.3	35.0	55.0	50.0	40.0	45.0	–	–	–	0.55
	12	60.0	38.3	38.3	41.7	42.5	37.5	–	–	–	
	16	48.3	24.2	36.7	35.0	58.3	34.2	–	–	–	
C	14	58.3	69.2	26.7	65.0	43.3	61.7	56.7	45.8	58.3	0.41
	18	36.7	28.3	55.0	65.0	45.0	46.7	62.5	58.3	41.7	

*Significance level

The above rationales uphold the notion that NASA–TLX scores are meaningful as the reference data by which the appropriateness of the TACOM measure can be established. For this reason, a linear regression analysis is conducted using the data summarized in Table 9.5. Figure 9.2 shows the results of a statistical analysis with ANOVA table.

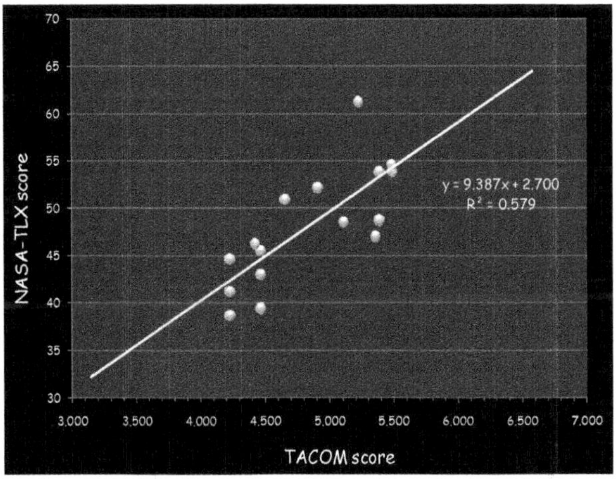

ANOVA table

Item	Degree of freedom	Sum of squares	Mean square	F statistics
Model	1	326.498	326.498	19.207
Error	14	237.982	16.999	
Total	15	564.480		

$F_{0.05}(1, 14) = 4.600$
$p < 10^{-4}$

Residual analysis

- Residual mean: -9.770×10^{-15}
- Normality test: passed ($p = 0.842$)
- Constant variance test: passed ($p = 0.512$)

Fig. 9.2 Result of linear regression analysis – TACOM scores with associated NASA–TLX scores

Figure 9.2 shows a remarkable correlation between TACOM scores and the associated NASA–TLX scores. In addition, the ANOVA table elucidates that the variation in NASA–TLX scores is largely attributable to the variation in TACOM scores ($p < 10^{-4}$). Therefore, it is reasonable to say that the TACOM measure is meaningful for explaining subjective workload scores perceived by SROs.

9.3 Comparing Task Performance Time Data Obtained from Other NPPs

In studying human-performance-related issues, one of the important findings is that the performance of qualified operators (or unqualified operators) is predictable when they are carrying out tasks having similar complexities (Chater 2000; Feldman 2000; Hamilton and Clarke 2005; Johannsen et al. 1994; Johnson and Payne 1985; Ogawa 1993; Stassen et al. 1990; Stanton and Young 1999; Zandin 2003). From the point of view of proceduralized tasks, one plausible explanation of this finding is that procedures strongly affect the actual behavior of qualified operators by institutionalizing detailed instructions. In other words, since proceduralized tasks institutionalize what is to be done and how to do it, it is assumed that the performance of qualified operators is, to some extent, predictable. Actually, the results of existing studies have provided a theoretical as well as an empirical clue supporting the reasonability of this assumption (Hollnagel et al. 1999; Kim et al. 2003; Stanton and Baber 2005).

If we adopt this assumption, it is natural to expect that the appropriateness of the TACOM measure can be consolidated by comparing TACOM scores with task performance time data gathered from other NPPs. For the sake of convenience, it should be noted that NPPs from which task performance time data were additionally collected will henceforth be referred as the subsidiary reference NPPs.

Similar to the case of the reference NPPs, a full-scope simulator has been installed in the training center of the subsidiary reference NPPs. This simulator is designed based on the MCR of a PWR that has 950 MWe capacity with conventional control devices. In addition, qualified operators working in the MCR of the subsidiary reference NPPs must be regularly retrained in order to increase their skills or knowledge related to various operating conditions including emergencies. Therefore, it is possible to collect audiovisual records on emergency operations under SGTR conditions that were carried out by 6 MCR operating teams. This collection was conducted from April to August 2005, and as a result, averaged task performance time data on 9 distinctive emergency tasks were obtained. Table 9.7 summarizes averaged performance time data on emergency tasks with their associated TACOM scores.

Based on the task performance time data shown in Table 9.7, a direct comparison was conducted to clarify whether averaged task performance time data obtained from the subsidiary reference NPPs remained within a certain range predicted by those from the reference NPPs. Figure 9.3 depicts the results of this comparison.

Table 9.7 Averaged task performance time data with the associated TACOM scores that are collected from the subsidiary reference NPPs (Park and Jung 2008, © Elsevier)

ID	TS	TR	TU	TACOM	Avg.(s)[1]	SD(s)[2]
1	4.626	1.774	4.112	4.051	41.9	25.5
2	4.630	1.496	3.495	3.944	12.0	2.9
3	4.042	1.821	3.979	3.627	17.9	5.6
4	4.691	1.799	4.262	4.121	33.9	22.3
5	5.486	2.203	4.134	4.716	55.4	27.8
6	4.847	1.680	3.879	4.168	38.9	16.0
7	4.433	1.537	3.778	3.843	34.7	10.3
8	5.976	2.740	6.344	5.441	97.0	28.6
9	5.742	2.547	5.227	5.084	77.1	24.1

[1]Avg.(s) denotes the mean value of task performance time data for each emergency task
[2]SD: standard deviation

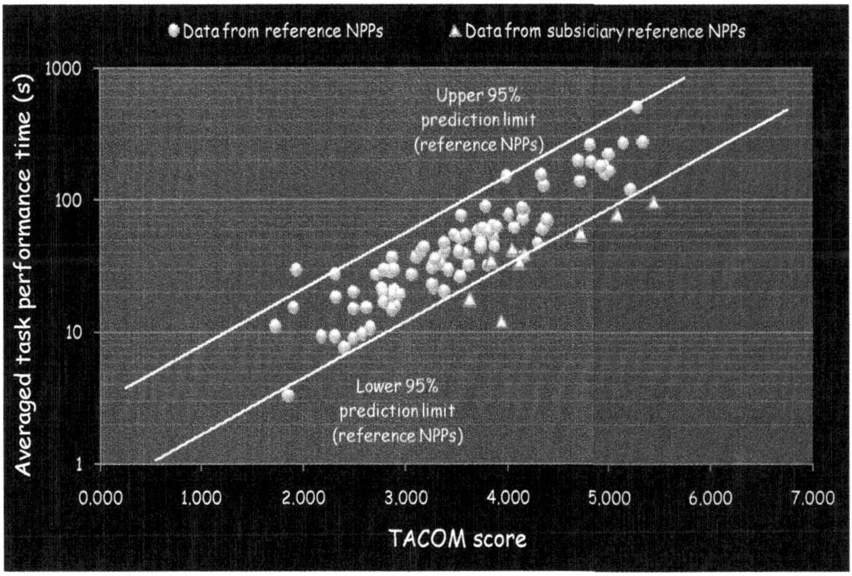

Fig. 9.3 Comparing two sets of task performance time data

In Fig. 9.3, there are two lines, *Upper 95% prediction limit* and *Lower 95% prediction limit*. Here, the meaning of the former is that, with a 95% confidence level, most of the averaged task performance time data obtained from the reference NPPs are expected to not exceed this limitation. Similarly, *Lower 95% prediction limit* indicates that, with a 95% confidence level, most of the averaged task performance time data will be greater than this limitation. Under these prediction limits, it is anticipated that two sets of task performance time data will be comparable with respect to TACOM scores because most of the task performance time data obtained from the subsidiary reference NPPs seem to be located near the lower prediction limit. In other words, although the contents of emergency tasks to be done by qualified operators working in the reference NPPs are quite different from those of the subsidiary reference NPPs, averaged task performance time data are predictable to some extent when the complexity score of a task (i.e., TACOM score) is given.

This expectation becomes more evident when averaged task performance time data obtained from the subsidiary reference NPPs are compared with those of the reference NPPs, which are obtained under similar conditions. Table 9.8 summarizes averaged task performance time data extracted from the OPERA database and collected under SGTR conditions of the reference NPPs. In addition, Fig. 9.4 depicts the results of these comparisons.

Table 9.8 Averaged task performance time data with the associated TACOM scores pertaining to the SGTR condition of the reference NPPs (Park and Jung 2008, © Elsevier)

ID	TS	TR	TU	TACOM	Avg.(s)	SD(s)
1	2.807	1.612	2.846	2.579	10.5	6.14
2	3.384	1.434	2.404	2.900	13.5	7.55
3	4.005	2.186	4.901	3.804	32.0	11.14
4	4.698	2.450	4.884	4.299	49.5	17.87
5	3.226	1.612	2.846	2.867	18.6	9.23
6	4.429	2.450	4.549	4.064	48.4	11.72
7	3.724	1.478	3.374	3.276	36.8	30.56
8	4.317	1.806	2.856	3.674	49.1	24.71
9	4.264	2.099	4.863	3.956	44.1	19.70
10	4.846	2.154	3.814	4.210	89.0	62.20
11	5.447	2.550	6.214	5.038	169	66.70
12	6.007	2.285	6.178	5.385	507	239.40

Figure 9.4 is very important for clarifying the appropriateness of the TACOM measure. According to Stassen et al. (1990), it was pointed out that human performance could be predictable if tasks are well defined. In addition, laboratory experiments have shown that the performance of human operators would be the

same if systems to be supervised had the same complexity, although the systems might differ in the number of functions and the degree of interactions (Wieringa and Stassen 1993). Therefore, the concept of an iso-complexity curve was suggested based on the number of functions and the degree of interactions (Johannsen et al. 1994; Visser and Wieringa 2001). This strongly suggests that, even though qualified operators have to accomplish different tasks, if there is a proper measure that can evaluate the complexity of a well-defined task, then their performance should not only be predictable but also be standardized as a function of a task complexity score. Subsequently, it is possible to say that the TACOM measure is meaningful for quantifying the complexity of a task to be done by qualified operators.

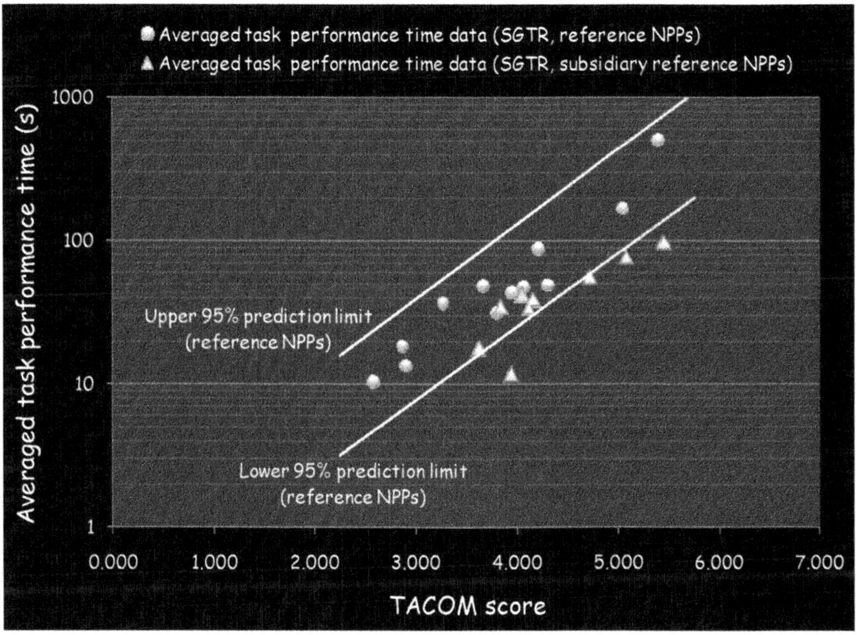

Fig. 9.4 Comparing two sets of averaged task performance time data collected under SGTR conditions

References

Bartko JJ (1966) The intraclass correlation coefficient as a measure of reliability. Psychol Rep 19:3–11

Bartko JJ (1976) On various intraclass correlation reliability coefficients. Psychol Bull 83:762–765

Bruton A, Conway JH, Holgate ST (2000) Reliability: what is it and how is it measured? Physiotherapy 86(2):94–99

Campbell DJ (1988) Task complexity: a review and analysis. Acad Manage Rev 13(1):40–52

Chater N (2000) The logic of human learning. Nature 407:572–573
Dickinson J, Byblow WD, Ryan LA (1993) Order effects and weighting process in workload assessment. Appl Ergonom 33(1):17–33
Feldman J (2000) Minimization of Boolean complexity in human concept learning. Nature 407:630–633
Hamilton WL, Clarke T (2005) Driver performance modelling and its practical application to railway safety. Appl Ergonom 36:661–670
Hancock PA (1996) Effects of control order, augmented feedback, input device and practice on tracking performance and perceived workload. Ergonomics 39(9):1146–1162
Hart SG, Staveland LE (1988) Development of NASA-TLX (Task Load Index): results of empirical and theoretical research. In: Hancock PA, Meshkati N (eds) Human Mental Workload, Elsevier, Amsterdam, pp.139–183
Hendy KC, Hamilton KM, Landry LN (1993) Measuring subjective workload: when is one scale better than many? Hum Factors 35(4):579–601
Henneman RL, Rouse WB (1984) Measures of human problem solving performance in fault diagnosis tasks. IEEE Trans Syst Man Cybern 14:99–112
Hill SG, Iavecchia HP, Byers JC, Bittner, Jr., AC, Zaklad AL, Christ RE (1992) Comparison of four subjective workload rating scales. Hum Factors 34(4):429–439
Hollnagel E, Kaarstad M, Lee HC (1999) Error mode prediction. Ergonomics 42:1457–1471
Johannsen G, Levis AH, Stassen HG (1994) Theoretical problems in man-machine systems and their experimental validation. Automatica 30:217–231
Johnson EJ, Payne JW (1985) Effort and accuracy in choice. Manage Sci 31:395–414
Kim JH, Lee SJ, Seong PH (2003) Investigation on applicability of information theory to prediction of operator performance in diagnosis tasks at nuclear power plants. IEEE Trans Nuclear Sci 50:1238–1252
Landis JR, Koch GG (1977) The measurement of observer agreement for categorical data. Biometrics 33:159–174
Levy AS, Lintner S, Kenter K, Speer KP (1999) Intra- and interobserver reproducibility of the shoulder laxity examination. Am J Sport Med 27(4):460–463
Li K, Wieringa PA (2000) Understanding perceived complexity in human supervisory control. Cognit Technol Work 2:75–88
Liu Y, Wickens CD (1994) Mental workload and cognitive task automaticity: an evaluation of subjective and time estimation metrics. Ergonom 37(11):1843–1854
Marinus J, Visser M, Stiggelbout AM, Rabey JM, Martinez-Martin P, Bonuccelli U, Kraus PH, Hilten JJ (2004) A short scale for the assessment of motor impairments and disabilities in Parkinson's disease: the SPES/SCOPA. J Neurol Neurosurg Psychiatr 75:388–395
Maynard DC, Hakel MD (1997) Effects of objective and subjective task complexity on performance. Hum Perform 10(4):303–330
Moray N (1999) Advanced displays, cultural stereotypes and organizational characteristics of a control room. In: Misumi J, Wilpert M, Miller R (eds) Nuclear Safety: A Human Factors Perspective. Taylor & Francis, New York
NASA (2009) http://humansystems.arc.nasa.gov/groups/TLX/
Nygren TE (1991) Psychometric properties of subjective workload measurement techniques: implications for their use in the assessment of perceived mental workload. Hum Factors 33(1):17–33
Ogawa K(1993) A complexity measure of task content in information-input tasks. Int J Hum-Comput Interact 5(2):167–188
Park J, Jung W (2006) A study on the validity of a task complexity measure for emergency operating procedures of nuclear power plants – comparing with a subjective workload. IEEE Trans Nuclear Sci 53(5):2962– 2970
Park J, Jung W (2008) A study on the validity of a task complexity measure for emergency operating procedures of nuclear power plants – comparing task complexity scores with two sets of operator response time data obtained under a simulated SGTR. Reliabil Eng Syst Saf

93:557–566

Reinartz SJ, Reinartz G (1992) Verbal communication in collective control of simulated nuclear power plant incidents. Reliabil Eng Syst Saf 36:245–251

Stanton N, Young M (1999) What price ergonomics? Nature 399:197–198

Stanton NA, Baber C (2005) Validating task analysis for error identification: reliability and validity of a human error prediction technique. Ergonomics 48:1097–1113

Stassen HG, Johannsen G, Moray N (1990) Internal representation, internal model, human performance model and mental workload. Automatica 26(4):811–820

Svensson E, Angelbrog-Thandrez M, Sjoberg L, Olsson S (1997) Information complexity: mental workload and performance in combat aircraft. Ergonomics 40:362–380

Vidulich MA, Tsang PS (1986) Technique of subjective workload assessment: a comparison of SWAT and the NASA-Bipolar methods. Ergonomics 29(11):1385–1398

Visser M, Wieringa PA (2001) PREHEP: Human error probability based process unit selection. IEEE Trans Syst Man Cybern C Appl Rev 31(1):1–15

Wei ZG, Macwan AP, Wieringa PA (1998) A quantitative measure for degree of automation and its relation to system performance and mental load. Hum Factors 40(2):277–295

Wieringa PA, Stassen HG (1993) Assessment of complexity. In: Wise JA, Hopkin VD, Stager P (eds) Verification and validation of complex systems: Human Factors Issues, Springer, Berlin, Heiddelberg, New York, pp.173–180

Zandin KB (2003) MOST Work Measurement Systems, 3rd edn. Marcel Dekker, New York

Part III
Promising Applications and Outlook

10 Promising Applications

As explained in the 6 chapters of Part II, the TACOM measure was developed to evaluate the complexity of proceduralized tasks by quantifying complexity factors pertaining to the performance of a process control task. To this end, each action to be performed by qualified operators has been analyzed from the point of view of an *OBJECT*, an *ACTION VERB,* and action specifications that can be subdivided into a *MEANS*, an *ACCEPTANCE CRITERION*, a *CONSTRAINT,* and a *peculiarity*. This strongly indicates that the TACOM measure is a verbatim probe evaluating the complexity of proceduralized tasks as written. In other words, the TACOM measure provides not a subjective but an objective value representing the *verbatim complexity* of proceduralized tasks that is to be loaded on qualified operators who have diverse individualities, such as aptitude, capability, cognitive style, motivation, self confidence, etc.

For example, washing both hands is a very easy task for many people. However, for some people, this task could be more complicated than it seems if they worried about the fact that many actions must be done simultaneously within a very short time: (1) turn on the water, (2) get soap, (3) rub soap on hands, (4) put the soap down, (5) rub both hands, (6) submerge both hands under water, (7) rub both hands, and (8) turn off the water. In an extreme case, someone might become more anxious about this task because the number of actions would vary from person to person. This means that the levels of a task's complexity felt by qualified operators would be widely dispersed, even though they performed the same task. Accordingly, it is very difficult to develop an effective strategy by which the countermeasures to reduce the possibility of human error (or to enhance the performance of qualified operators) can be identified. However, since the TACOM measure quantifies the complexity of proceduralized tasks based on a task description, it is reasonable to expect that useful guidelines or insights to support qualified operators can be identified from an analysis of TACOM scores.

10.1 Providing HRA Inputs

From the point of view of engineering, the most popular approach to coping with human error is to develop a method that can be used not only to quantify the possibility of human error but also to identify crucial factors causing human error.

This approach is widely known as HRA (human reliability analysis or human reliability assessment). In order to conduct HRA, many kinds of information should be provided to HRA practitioners. Typical information includes the following (Cooper et al. 1996; Hollnagel 1993b; IAEA 1990; IEEE 1997; Kirwan 1994; Kirwan and Ainsworth 1992; Sträter and Bubb 1999; Swain and Guttmann 1983):

- Description of the tasks to be performed
- List of available (or to be used) procedures
- The experience level of qualified operators (or teams) who have to perform the required tasks
- The dependence among the required tasks
- An allowable time window by which the required tasks should be completed
- The time needed to perform the required tasks (i.e., task performance time)

Of these, time-related information (i.e., the available time as well as the task performance time) is essential. Briefly, the available time is the difference between an allowable time and a task performance time, as illustrated in Fig. 10.1.

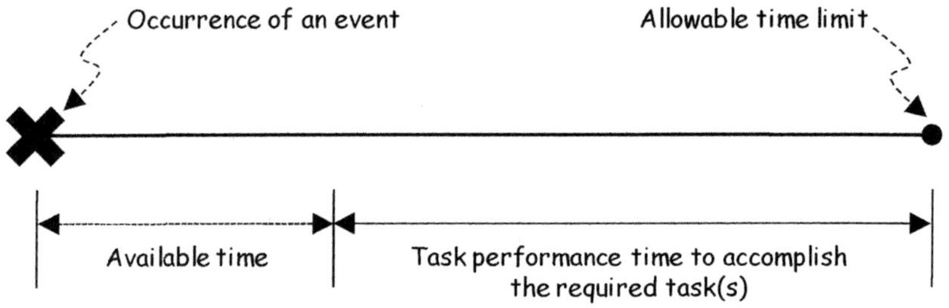

Fig. 10.1 Allowable time, task performance time, and available time

For example, when an SGTR has occurred, it is strongly recommended that qualified operators should successfully isolate a ruptured SG within about 30 min by following a set of proceduralized tasks described in an SGTR procedure. In this case, if qualified operators need at least 20 min to complete the required tasks, then 10 min are available to correctly recognize the occurrence of the SGTR. This implies that qualified operators are likely to make a mistake in recognizing the occurrence as well as the nature of an ongoing situation because 10 min does not seem to be enough time. In addition, if qualified operators fail to recognize the situation within 10 min, then they are apt to make an additional mistake in the course of performing the required tasks because they have to accomplish what should be done more quickly (i.e., time pressure). Accordingly, the possibility of human error increases as the decrease of the available time (Hollnagel 1993b; Kozine 2007; Woods et al. 1984; Williams 1988).

Here, since the allowable time can be estimated by deterministic approaches (e.g., a thermohydraulic experiment or a theoretical analysis), it is possible to say that the available time is a function of the task performance time. However, it is very difficult to gather a sufficient amount of task performance time data based on operating experience because of the infrequency of occurrence of an emergency event. For this reason, although several divergences from a real-life situation (i.e., a fidelity problem) still make it possible to dispute the use of simulators (Stanton 1996; O'Hara and Hall 1992; Hollnagel 2000; IAEA 2004), it is apparent that the use of simulators has been regarded as the most cost- and effort-effective way in collecting task performance time data, especially in an emergency situation (Stanton 1996; Rasmussen and Jensen 1974; IAEA 2004). Nevertheless, the use of simulators is still problematic, because a huge amount of resources (e.g., manpower, time, and cost) is generally required to simulate emergency events.

In light of these concerns, the TACOM measure seems to be a practicable solution because there is a strong correlation between TACOM scores and task performance time data. That is, as depicted in Figs. 8.7 and 9.3, the TACOM measure should be able to estimate task performance time data with an upper as well as a lower prediction limit when the TACOM scores of the required tasks are given. Actually, Chi and Chung (1996) and Hamilton and Clarke (2005) have independently shown that task performance time data predicted by a theoretical model are directly comparable to those which are actually observed. This means that, from the point of view of HRA, estimating the possibility of human error based on the predicted task performance time (or the available time) is a viable approach.

However, although HRA is a useful tool to cope with human errors, a more straightforward way would be the management of complicated tasks that challenge the cognitive ability of qualified operators. That is, if we recall that a significant portion of human error are caused by complicated tasks that force qualified operators to use a lot of cognitive resources exceeding their cognitive ability, the identification of complicated tasks that are likely to place an excessive workload on qualified operators seems to be indispensable.

10.2 Identifying Complicated Tasks Demanding an Excessive Workload

As stated at the end of Chap. 2, the complexity of proceduralized tasks should be managed because the complexity increases the possibility of human error by placing an excessive workload on qualified operators. Accordingly, we at least have to answer one crucial question – *how can we identify a complicated task demanding an excessive workload of qualified operators*?

In this regard, it is very interesting to point out that a complicated task increases the possibility of violations by making qualified operators look for more effective shortcuts. That is, as depicted in Fig. 2.4, qualified operators are likely to deviate from a procedure if they believe that there is a better way to accomplish a complicated task demanding an undue workload. Therefore, scrutinizing the characteristics of procedure deviations along with changes in TACOM scores would provide us with an important clue regarding the identification of complicated tasks. For this reason, the behavior types of SROs who must shoulder most of the burden arising from the performance of emergency tasks are worth investigating.

10.2.1 Three Kinds of Behavior Types in Conducting Procedural Steps

The audiovisual records of retraining sessions, which were the data sources of the OPERA database, have been meticulously analyzed in order to observe how SROs have carried out emergency tasks included in EOPs. In particular, these observations have focused on the performance of procedural steps because they are the minimal unit of emergency tasks (i.e., each emergency task consists of one or more procedural steps). Consequently, as summarized in Table 10.1, three types of distinctive behaviors are identified from SROs' activities.

Table 10.1 SROs' behaviors pertaining to the performance of procedural steps included in EOPs

Type		Meaning
A	Strict adherence	SROs strictly follow all the required actions as written
B	Skipping redundant actions	SROs skip an action that is identical to one that already carried out in the previous procedural step
		SROs perform the same action based on previously known information
C	Modifying the sequence of actions	SROs carry out a procedural step using a modified sequence of actions that is different from the predefined sequence of actions

From Table 10.1, *Type A (strict adherence)* means that SROs have conducted all the required actions along with the predefined sequence of actions (i.e. compliance behavior). In contrast, both *Type B (skipping redundant actions)* and *Type C (modifying the sequence of actions)* imply typical noncompliance behaviors related to finding an effective shortcut. In order to understand the characteristics of noncompliance behaviors, let us consider Fig. 10.2, which shows three arbitrary procedural steps included in EOPs.

First, *Type B* denotes that SROs conduct all the required actions included in a procedural step to be performed, excluding redundant actions that were already conducted in the previous procedural step (i.e., prior actions). For example, as can be seen from Fig. 10.2, *verify containment pressure is less than 70* cmH_2O action is commonly included in both *Steps 1* and *2*. In this case, it has been frequently observed that SROs did not check the current value of containment pressure in the course of performing *Step 2*, since they already checked it in *Step 1*. In addition, instead of skipping this action, several SROs performed this action by themselves (i.e., without communicating with board operators) based on the old value of the containment pressure obtained in the course of performing *Step 1*.

	Instructions	Contingency Actions
Step 1	Determine the containment isolation acceptance criteria are met by performing **ALL** of the following: a. Verify containment pressure is less than 70 cmH_2O. b. Verify **NO** containment area radiation alarms or unexplained rise in radiation has occurred. c. Verify **NO** steam plant radiation alarms or unexplained rise in radiation has occurred.	a. **IF** containment pressure is greater than 133.1 cmH_2O, **THEN** ensure CIAS is actuated. b. **IF** there is steam plant radiation alarm or unexplained rise in radiation, **THEN** sample SG activity.
Step 2	Determine containment temperature and pressure acceptance criteria are met by performing **BOTH** of the following: a. Verify containment temperature is less than 49°C. b. Verify containment pressure is less than 70cmH_2O.	a. Ensure all required containment normal cooling and ventilation systems are in operation: *... (rest of actions)*
Step 3	**IF** containment pressure is greater than 1423.6 kg/cm^2, **THEN** perform **ALL** of the following: a. Verify CSAS (containment spray actuation signal) is actuated automatically. b. Verify all CS (containment spray) pumps are delivering at least 15,200 LPM (liter per minute) c. Close RCP (reactor coolant pump) seal leak-off isolation valves. d. Stop all RCPs.	a. **IF** CSAS has **NOT** been initiated automatically **THEN** manually actuate CSAS. • EF-HS-101A/101B/101C/101D. b. **IF ANY** CS pumps **CANNOT** deliver 15,200 LPM **THEN** perform **ANY** of the following: *... (rest of actions)*

Fig. 10.2 Three arbitrary procedural steps to explain *Type B* and *Type C* behavior (Park and Jung 2003, © Elsevier)

Second, *Type C* indicates that SROs carry out the required actions based on a modified sequence of actions. It has been frequently observed that SROs seem to try to change the predefined sequence of actions into another one in order to perform a procedural step more easily. It is to be noted that the main difference between *Type B* and *Type C* is the existence of *prior actions*, since *Type C* automatically includes the behavior of skipping actions due to the modified sequence of actions. Let us consider Fig. 10.3, which depicts the ACG of *Step 3*.

First, when SROs start to perform *Step 3*, they have to verify whether the containment pressure is greater than 1423.6 kg/cm^2 or not (refer to the first action in Fig.10.3). If the result is *yes*, then SROs have to perform either *verify all containment spray (CS) pumps are delivering at least 15200 LPM* action or *manually actuate containment spray actuation signal (CSAS)* action based on the results of *verify CSAS is actuated automatically* action. However, several SROs accomplished this procedural step using a modified action sequence, as illustrated in Fig. 10.4.

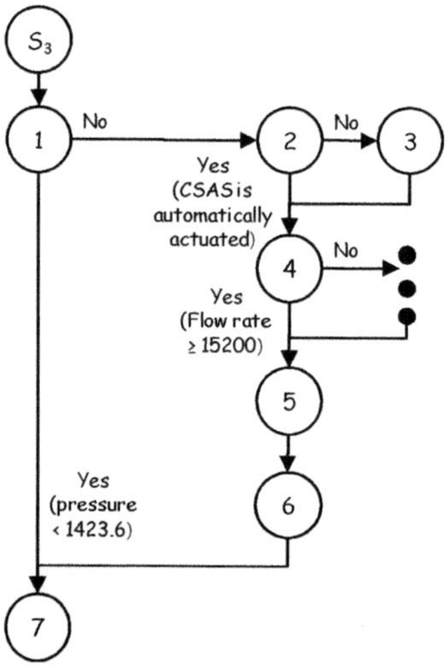

ID	Action description
S_3	Perform Step 3
1	Verify containment pressure is greater than 1423.6 kg/cm^2
2	Verify CSAS is actuated automatically
3	Manually actuate CSAS
4	Verify all CS pumps are delivering at least 15200 LPM
5	Close RCP seal leak-off isolation valves
6	Stop all RCPs
7	Go to the next procedural step

Fig. 10.3 ACG of *Step 3* (Park and Jung 2003, © Elsevier)

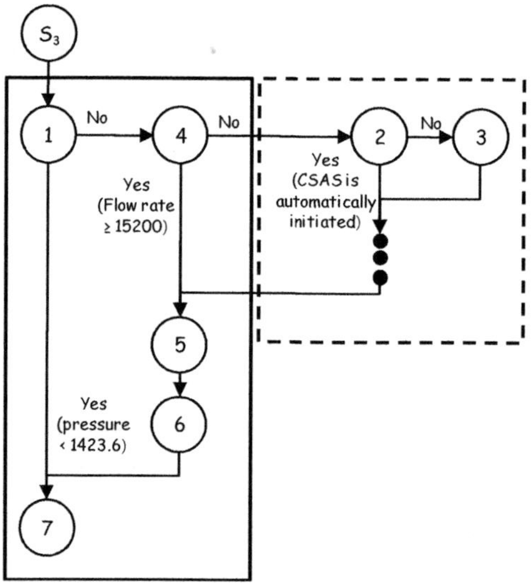

Fig. 10.4 Modified sequence of actions about *Step 3* (Park and Jung 2003, © Elsevier)

As shown in Fig. 10.4, SROs carried out *verify all CS pumps are delivering at least 15200 LPM* action before conducting *verify CSAS is actuated automatically* action. This sequence of actions is the deviation from the predefined one depicted in Fig. 10.3. Nevertheless, the fruit of this modification seems to be attractive – *reducing the number of actions to be conducted by SROs*. This is because SROs do not need to consider the several actions enclosed by dotted lines when the flow rate of CS pumps is greater than 15200 LPM.

From the above examples, thus, the meaning of prior actions could become obvious, since the only way to discriminate *Type B* from *Type C* is to check the existence of identical actions. It is to be noted that there will be many different types of noncompliance behaviors that can be observed in the course of performing procedural steps. Unfortunately, it is very difficult to detect other types of noncompliance behaviors because most of them have occurred in the mental processes of SROs. Accordingly, for the sake of simplicity, it is assumed that all the noncompliance behaviors belong to either *Type B* or *Type C*.

10.2.2 The Meaning of Noncompliance Behaviors

It seems that there is a plausible explanation why SROs adopt these types of noncompliance behaviors. As one of the training instructors working in the reference NPPs stated:

> When the containment pressure is high, SROs ultimately want to know whether a sufficient CS flow is delivered or not. In addition, most SROs already recognize that, when the CSAS is actuated, CS pumps and the associated valves are automatically aligned in order to deliver sufficient CS flow. Thus, the adoption of Type C is understandable, because they are able to reduce the number of the required actions by checking flow rate from CS pumps before anything else.

At the same time, however, the training instructor also noted that both *Type B* and *Type C* might be risky, because these noncompliance behaviors can directly result in an unanticipated consequence. For example, licensee event reports (LERs) issued in the U.S.A have revealed that a significant portion of incidents was caused by a noncompliance behavior such as *an operator's decision upon a course of action based on what information he had* (Brune and Weinstein 1981). In addition, Macwan and Mosleh (1994) stated that *memory of recent actions* is one of the causes resulting in a procedure-related human error. That is, when qualified operators are asked to verify the flow rate, they are apt to omit verifying the current value of the flow if they have recently verified that the status of the associated pump is running.

Nevertheless, the above explanations clearly show that both *Type B* and *Type C* are not malicious but a kind of optimized response to satisfactorily perform the required tasks under a given constraint. This means that the comparison between noncompliance behaviors and TACOM scores would be meaningful because qualified operators will try to reduce the amount of undue workload by adopting a more effective way to perform procedural steps.

10.2.3 Comparing the Occurrence of Noncompliance Behaviors with the Associated TACOM Scores

In order to compare noncompliance behaviors with the associated TACOM scores, the OPERA database has been meticulously examined. As a result, Table 10.2 summarizes a profile about the number of compliance as well as noncompliance behaviors, which is grouped so that the distribution of observations is fit to a normal distribution with respect to TACOM scores (Kolmogorov-Smirnov test passed, $p > 0.2$).

Table 10.2 Profile of compliance as well as noncompliance behaviors

TACOM score	Number of observations			
(bin size = 0.6)	Type A	Type B	Type C	Total
1.401 ~ 2.000	28	0	1	29
2.001 ~ 2.600	143	20	37	200
2.601 ~ 3.200	332	32	139	503
3.201 ~ 3.800	175	3	55	233
3.801 ~ 4.400	104	7	19	130

In order to clarify whether the occurrences of noncompliance behaviors are influenced by the associated TACOM scores, the χ^2 test is conducted as summarized in Table 10.3.

Table 10.3 Results of χ^2 test

TACOM score		The number of observations			The number of expectations		
Range	Representative	Type A	Type B	Type C	Type A	Type B	Type C
1.401 ~ 2.000	1.700	28	0	1	20.7	1.6	6.6
2.001 ~ 2.600	2.300	143	20	37	142.8	11.3	45.8
2.601 ~ 3.200	2.900	332	32	139	359.2	28.5	115.3
3.201 ~ 3.800	3.500	175	3	55	166.4	13.2	53.4
3.801 ~ 4.400	4.100	104	7	19	92.8	7.4	29.8

$\chi^2 = 38.4$, df (degrees of freedom) = 8, $p < 10^{-3}$; rejection criterion = $\chi^2_{0.05}(8) = 15.5$

As a result, it seems that the occurrences of compliance behaviors are able to be explained by TACOM scores since the χ^2 value is greater than the rejection criterion for the null hypothesis (e.g., $\chi^2 = 32.1 > \chi^2_{0.05}(8) = 15.5$). This means that qualified operators are likely to change their behaviors with respect to the complexity of procedural steps. If we adopt this expectation, then it is meaningful to compare the effect of TACOM scores on the percentage of compliance behaviors (Fig. 10.5).

From Fig. 10.5, it is observed that many SROs seem to adopt noncompliance behaviors more frequently when they have to conduct procedural steps whose TACOM scores range from 2.300 to 3.500 (based on representative values). In contrast, when SROs are faced with procedural steps whose TACOM scores are either relatively low (i.e., less than 2.300) or relatively high (i.e., greater than 3.500), they seem to try to follow procedural steps as written.

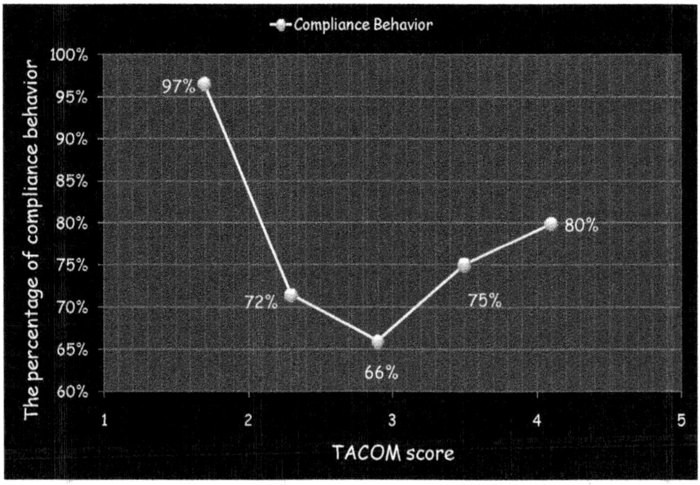

Fig. 10.5 Comparing the percentage of compliance behaviors with the associated TACOM scores

10.2.4 Criterion for Complicated Tasks

As can be seen from Fig. 10.5, the relation between compliance behaviors and TACOM scores shows a large U shape (or an inverted-U shape for noncompliance behaviors). In this regard, it is possible to assume that we are able to establish a criterion for complicated procedural steps demanding an excessive workload. To this end, let us consider Fig. 10.6.

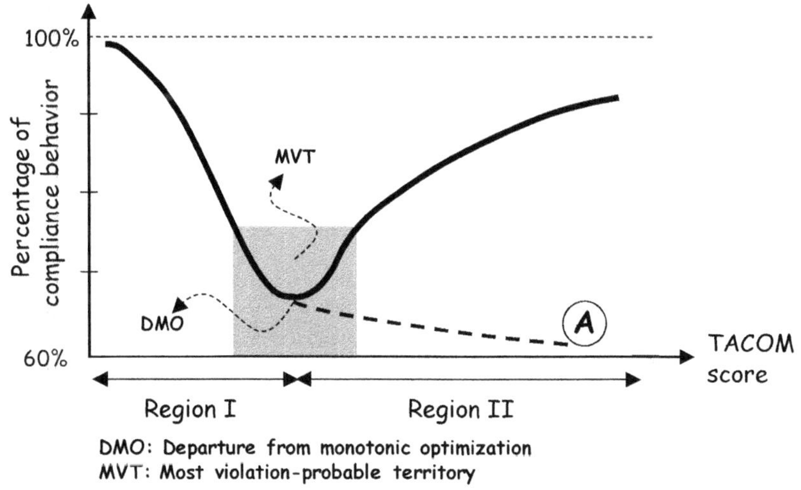

Fig. 10.6 Hypothetical tendency of compliance behaviors with respect to an increase in TACOM scores

In *Region I*, SROs show an expected tendency to frequently adopt noncompliance behaviors (i.e., searching for a shortcut) accompanied by an increase in TACOM scores. If this tendency continues, the percentage of noncompliance behaviors will follow a hypothetical line that is monotonically falling such as Ⓐ in Fig. 10.6. However, in *Region II*, observed data show that SROs seem to less frequently adopt noncompliance behaviors when they exceed a certain value of the TACOM measure. In other words, SROs seem to try to carry out the required actions as written even if they have to accomplish more complicated procedural steps. This contradictory tendency can be understood if we consider two assumptions from the point of view of optimization behavior.

First, when SROs are faced with a procedural step that consists of a few actions with a simple action sequence, they will likely to carry it out as written. This is because the procedural step is so easy that SROs do not need to consider noncompliance behaviors to reduce an undue workload. Meanwhile, in the case of a complicated procedural step, it is assumed that SROs might feel a burden in adopting noncompliance behaviors because there is no benefit to reducing an undue workload. That is, customizing a complicated procedural step through adopting noncompliance behaviors is not favorable since SROs may use a considerable amount of cognitive resources dealing with various kinds of causalities, such as the automatic running of CS pumps due to the actuation of the CSAS, in the course of searching for a shortcut. For this reason, the inflection point from which the percentage of compliance behaviors starts to increase can be referred to as the departure from monotonic optimization (DMO). According to Fig. 10.5, in the case of qualified operators working in the reference NPPs, it is expected that the DMO will be located somewhere in the range 2.300 to 3.500. Here, we are able to refer to this territory as the most violation-probable territory (MVT), because the chance of an unintended violation is relatively high in an unstable environment.

Fortunately, these assumptions appear to be reasonable because it is anticipated that SROs will just try to trade off noncompliance behaviors with the complexity of procedural steps (i.e., cost-benefit trade-offs) (Reason 2008). For example, Amalberti (2001) pointed out that "Fundamentally, an operator does not regulate the risk of error, he regulates a high performance objective at the lowest possible execution cost. In the human mind, error is a necessary component of this optimized performance result (p. 118)." Similarly, Leplat (1998) stated that "These studies, for example, have shown that when the demands or the complexity of the work increase, one process for reducing complexity is to change work method (p. 110)."

And Vicente (1999) explained:

> At one plant, operators would not always follow the written procedures when they went to the simulator for recertification. They deviated from them for one of two reasons. In some cases, operators achieved the same goal using a different, but equally safe and efficient, set of actions. … In other cases, the operators would deviate from the procedures because the desired goal would not be achieved if the procedures were followed. It is very difficult to write a procedure to encompass all possible situations (p. xiii).

Therefore, the percentage of noncompliance behaviors will be proportional to the amount of benefits that are seen as outweighing the possible costs if SROs believe that they will not result in bad consequences (Dien et al. 1992; Maurino et al. 1995; Vessey 1994; Visciola et al. 1992; Lawton 1998). This strongly suggests that SROs are apt to adopt noncompliance behaviors when they have to perform procedural steps whose complexity is within a certain tolerable range. Subsequently, it is presumed that qualified operators are able to accomplish procedural steps whose TACOM scores are less than the DMO, with an acceptable workload. In contrast, qualified operators are likely to feel an excessive workload when they have to accomplish procedural steps whose TACOM scores are greater than the DMO. Here, if we assume that the value of the DMO is the best representative value of the MVT (i.e., 3.500), then 4.100 (i.e., the central value between 3.801 and 4.400) should be a representative value distinguishing a procedural step that might place an excessive workload on SROs. Consequently, it is highly expected that the possibility of procedure-related human errors (i.e., distraction-due-to-workload) will increase when qualified operators need to accomplish a proceduralized task that consists of a series of procedural steps whose TACOM scores are greater than this value. This implies that we might have a decisive clue for answering one of the pending issues in cognitive engineering: *In many hazardous technologies, the important issue is not whether to violate but when to violate* (see p. 291 of Reason et al. 1998).

Although a great amount of additional effort should be spent in advance to justify the aforementioned assumptions and expectations, it is hoped that the TACOM measure would contribute greatly to the identification of effective countermeasures to support qualified operators if we are able to establish a firm criterion regarding a complicated proceduralized task. In this vein, one of the typical contributions will be the provision of necessary inputs in the early phases of a human-machine interface (HMI) design process.

10.3 Providing Design Inputs on Effective HMIs

In general, it has been widely recognized that one of the key processes in the design of HMIs is task analysis. For example, as stated by Kirwan and Ainsworth (1992):

> Task analysis involves the study of what an operator (or team of operators) is required to do to achieve a system goal. The primary purpose of task analysis is to compare the demands of the system on the operator with the capabilities of the operator, and if necessary, to alter those demands, thereby reducing error and achieving successful performance (p. 15).

To this end, at least, it is essential to identify what kinds of information and activities are necessary to achieve the required tasks (Kirwan 1994; Kirwan and Ainsworth 1992; IEEE 1997). In this regard, Fig. 10.7 shows the typical results of a task analysis about the HMI design of NPPs (Lee et al. 1994).

Function	Regulating RCS inventory
Task	Increasing the rate of charging flow
Purpose	Increasing the rate of charging flow in order to compensate for expected condensations due to the cooling of RCS
Action	1. Switch the controller of charging flow to manual position. 2. Control the rate of charging flow until the water level of pressurizer reaches 70%. 3. If necessary, close BG-HV-1 and BG-HV-2. 4. Control the rate of charging flow less than 27 m^3/h. 5. If necessary, stop all remaining RCPs except one.
Indicator	1. CVCS (chemical and volume control system) charging flow indicator • BG-FI-122 (0-50m^3/h) 2. Pressurizer level: indicators • BB-LI-459A (0-100%) • BB-LI-460 (0-100%) • BB-LI-461 (0-100%) 3. Pressurizer level: trend recorder • LR-459 (0-100%)
Controller	1. CVCS charging flow controller • BG-FK-122 (manual: 0-100%, modulate) 2. CVCS letdown orifice valve switches • BG-HS-1 (Open, Close) • BG-HS-2 (Open, Close) 3. RCP controllers (Start, Stop)

Fig. 10.7 Typical results of a task analysis

It should be emphasized that the TACOM score of a task being considered can be directly quantified from the results of a task analysis because Fig. 10.7 contains all kinds of information for quantifying the five submeasures. This implies that more detailed as well as helpful functional specifications can be extracted in the early stages of an HMI design process. For example, Table 10.4 summarizes the TACOM score of *increasing the rate of charging flow* task.

Table 10.4 TACOM score of *increasing the rate of charging flow* task

SIC	SLC	SSC	AHC	EDC	TS	TR	TU	TACOM
3.640	2.000	3.000	4.564	4.736	3.458	2.279	4.736	3.436

It is to be noted that this task seems to be violation-probable, because the TACOM score shown in Table 10.4 belongs to a range in which qualified operators might adopt a noncompliance behavior more frequently. Due to this concern, we have to do something to reduce the possibility of an unintended violation about this task. Fortunately, the scores of the five submeasures provide diagnostic information by which an appropriate countermeasure can be figured out.

For example, it is anticipated that the EDC will be a primary contributor since its score is greater than those of the other submeasures. Actually, this anticipation seems to be reasonable because qualified operators have to conduct a couple of equally acceptable actions, such as *If necessary, close BG-HV-1 and BG-HV-2*, or *If necessary, stop all the remaining RCPs except one*. This means that it is indispensable for qualified operators to additionally provide either clearer task descriptions or more helpful information to support the selection of a proper action. However, as already explained at the end of Sect. 5.4, it is very difficult (or almost impossible) to describe detailed actions that accurately cover every situation it would be better to come up with the design of effective HMIs that provide supportive information to qualified operators. From this standpoint, it is expected that the TACOM measure can contribute to the design of effective HMIs in the following ways.

10.3.1 Clarifying the Types of Information Displays

The results of existing studies have revealed that the performance of qualified operators vary dramatically varied with respect to the appropriateness of information displays (Bennett et al. 1997; Goodstein 1981; Ham and Yoon 2001; Ham et al. 2008; Vicente 1999; Vicente and Rasmussen 1990; Wickens 1992; Woods 1991). In short, conventional information displays seem to be inappropriate for supporting the completion of required tasks that demand a high level of cognitive activities, such as searching for necessary information, interpreting information, and inferring information, etc. As a result, conventional information displays are likely to put a great cognitive burden on qualified operators who are working in a large and safety-critical process control system. Therefore, the provision of effective information displays is very important for enhancing the performance of qualified operators as well as, to some extent, for reducing the possibility of human errors.

In this regard, one of the essential questions is *what types of information displays are necessary to provide supportive information*? In other words, we need to clarify what kind of task-related information is necessary to decrease the amount of cognitive burden (or workload) to be placed on qualified operators. From this point of view, Vicente and Rasmussen (1992) suggested the framework of an ecological interface design (EID). Ham et al. (2008) summarized the features of the EID framework as follows:

> EID aims to systematically represent the identified work domain constraints in displays in order to support the adaptive, goal-directed human behavior. Two most important ingredients of the EID approach are identifying invariant constraints of work domains by employing AH (abstraction hierarchy) and designing information display to capitalize the human's powerful pattern recognition ability. The use of AH, a multilevel knowledge representation framework for describing the goal–means structure of work domains, allows designers to build a work domain model that makes human operators have a right mental model of the work domain. Up to now, there have been several studies proving the validity and effectiveness of the EID framework in diverse work domains. Collectively,

> these studies claimed that EID could lead to better performance than traditional displays. Cognitively complex tasks seemed to be more benefited from EID, compared to simple tasks; however, there were no harmful effects of EID under simple tasks (p. 255).

Here, it should be emphasized that the EID framework is effective for *cognitively complex tasks*. This strongly implies that the application of the EID framework should be selective for complicated tasks, because considerable time and effort are necessary to appy the EID framework to a large-scale problem (Vicente 2002). That is, in order to practically apply the EID framework to a large and safety-critical process control system, it should be combined with a kind of additional framework that can identify a complicated task challenging the cognitive ability of qualified operators (Jenkins et al. 2009). From this concern, it is expected that the TACOM measure could play an important role, because TACOM scores can identify complicated tasks that are likely to place an excessive workload on qualified operators. Consequently, one could say that the concurrent use of both the EID framework and the TACOM measure is a very promising approach to providing effective information displays.

10.3.2 Specifying Information Requirements for CBPs

From the point of view of providing supportive information, the use of a systematic framework to determine proper information displays in the early stages of an HMI design is an ideal solution. For example, the EID framework can be applied in the early stages of HMI design processes if a list of complicated tasks could be identified from the results of a task analysis. However, this solution is only available to a system to be constructed or being constructed. This means that we are able to come up with an alternative solution that can be applied to an operating system, such as NPPs. In this regard, a plausible solution would be to use a computer-based procedure (CBP), which is comparable to a paper-based procedure (PBP).

O'Hara et al. (2002) summarized the characteristics of both PBPs and CBPs as follows:

> PBPs also impose tasks on the operator that are not directly related to controlling the plant. To make transitions between procedure steps and documents, and maintain awareness of the status of procedures that are in progress, operators must handle, arrange, scan, and read PBPs in parallel with monitoring and control tasks. CBPs are being developed to support procedure management. CBPs have a range of capabilities that may support operators in controlling the plant and reduce the demands associated with PBPs. In the simplest form, CBPs show the same information via computer-driven video display units (VDUs). More advanced CBPs may include features to support managing procedures (e.g., making transitions between steps and documents, and maintaining awareness of procedures in progress), detecting and monitoring the plant's state and parameters, interpreting its status, and selecting actions and executing them (p. 1-1).

In sum, static PBPs have inherent drawbacks in supporting transitions among multiple procedures as well as a high level of cognitive activities that will dynamically vary with respect to an ongoing situation (such as interpreting process information or selecting appropriate actions). Therefore, CBPs have been developed for not only new NPPs but also existing NPPs with advanced computer and information technologies (Jung et al. 2004; Kontogiannis 1999a; Lipner and Kerch 1994; Pirus and Chambon 1997; Reynes and Beltranda 1990; Spurgin et al. 1988; Spurgin et al. 1993).

However, CBPs have not been widely used as expected because (1) there are still many unresolved issues and (2) practical guidance for their design is still insufficient (Kontogiannis 1999a; O'Hara et al. 2002; Niwa et al. 1996; Niwa and Hollnagel 2002). For example, one of the important design issues is the provision of supportive information to reduce *general cognitive workload* resulting from the high demand of cognitive activities, such as monitoring or decision making (O'Hara et al. 2002). Unfortunately, instead of practical guidelines that allow the designer of CBPs to identify what kind of information should be provided, only a list of high-level functional requirements is currently available.

In this regard, it is expected that another contribution of the TACOM measure could be the specification of design requirements for CBPs. In order to clarify this expectation, let us recall *verify the water level of Tank 1 is abnormally decreasing* action. As explained in Sect. 7.5, qualified operators probably need to check the water level of Tank 1 in parallel with the status of surrounding components to find out whether there is a good explanation for the decrease in the water level. If there is no evident cause, then qualified operators will suspect an abnormal decrease due to other factors, such as a break in a pipe. This implies that CBPs should support qualified operators by providing additional information, such as the status of related components or equipment, which is helpful for reducing the amount of cognitive resources to deal with an action description including an ambiguous *ACCEPTANCE CRITERION* (i.e., abnormally decreasing). Similarly, in the case of *align all the valves to transfer a coolant from Tank A to Tank B* action, CBPs should support qualified operators by providing the associated valves that are necessary to make a flow line to two tanks, because there is no specification about *MEANS*. In this way, it is possible to systematically articulate information requirements for CBPs.

References

Amalberti R (2001) The paradoxes of almost totally safe transportation systems. Saf Sci 37:109–126

Bennett KB, Nagy AL, Flach JM (1997) Visual displays. In: Salvendy G (ed) Handbook of Human Factors and Ergonomics, Wiley, New York, pp.659–696

Brune RL, Weinstein M (1981) Development of a checklist for evaluating emergency operating procedures used in nuclear power plants. NUREG/CR-1970, Washington, DC

Chi C, Chung K (1996) Task analysis for computer-aided design (CAD) at a keystroke level. Appl Ergonom 27(4): 255–265

Cooper SE, Ramey-Smith AM, Wreathall J, Parry GW, Bley DC, Luckas WJ, Taylor JH, Barriere MT (1996) A technique for human error analysis (ATHEANA) – technical basis and method description. NUREG/CR-6350, Washington, DC

Dien Y, Llory M, Montmayeul R (1992) Operators' knowledge, skill and know-how during the use of emergency procedures: design, training and cultural aspects. In: Proceedings of the 5th IEEE Conference on Human Factors in Power Plants, Monterey, CA, pp.178–181

Goodstein LP (1981) Discriminative display support for process operators. In: Rasmussen J, Rouse WB (eds) Human Detection and Diagnosis of System Failure, Plenum, New York, pp.433–449

Ham DH, Yoon WC (2001) The effects of presenting functionality abstracted information in fault diagnosis tasks. Reliabil Eng Syst Saf 73:103–119

Ham DH, Yoon WC, Han BT (2008) Experimental study on the effects of visualized functionality abstracted information on process control tasks. Reliabil Eng Syst Saf 93:254–270

Hamilton WL, Clarke T (2005) Driver performance modelling and its practical application to railway safety. Appl Ergonom 36:661–670

Hollnagel E (1993b) Human Reliability Analysis: Context and Control. Academic, London

Hollnagel E (2000) Looking for errors of omission and commission or the hunting of the Snark revisited. Reliabil Eng Syst Saf 68:135–145

IEEE (1997) IEEE guide for incorporating human action reliability analysis for nuclear power generating stations. IEEE Std 1082-1997, Piscataway

IAEA (1990) Human error classification and data collection. IAEA-TECDOC-538, Vienna

IAEA (2004) Use of control room simulators for training of nuclear power plant personnel. IAEA-TECDOC-1411, Vienna

Jenkins DP, Stanton NA, Salmon PM, Walker GH (2009) Cognitive Work Analysis: Coping with Complexity. Ashgate, London

Jung Y, Seong PH, Kim MC (2004) A model for computerized procedures based on flowcharts and success logic tree. Reliabil Eng Syst Saf 26:351–362

Kirwan B (1994) A guide to Practical Human Reliability Assessment. Taylor and Francis, London

Kirwan B, Ainsworth LK (1992) A Guide to Task Analysis. Taylor and Francis, London

Kontogiannis T (1999a) Applying information technology to the presentation of emergency operating procedures: implication for usability criteria. Behav Inf Technol 18(4):261–276

Kozine I (2007) Simulation of human performance in time-pressured scenarios. J Risk Reliabil 221:141–151

Lawton R (1998) Not working to rule: understanding procedural violations at work. Saf Sci 28(2):77–95

Lee YH, Cheon SW, Suh SM, Lee JW (1994) A survey on the task analysis methods and techniques for nuclear power plants operators. KAERI/AR-402, Daejeon, South Korea

Lee JW, Park JC, Lee YH, Oh IS, Lee HC, Jang TI, Kim DH, Hwang SH, Park JK, Kim JS (2008) Development of the digital reactor safety system. KAERI/RR-2909, Daejeon, South Korea

Leplat J (1998) Task complexity in work situations. In: Goodstein LP, Anderson HB, Olsen SE (eds) Tasks, Errors and Mental Models, Taylor and Francis, London, pp.105–115

Lipner MH, Kerch SP (1994) Operational benefits of an advanced computerized procedures system. In: Nuclear Science Symposium and Medical Imaging Conference, Norfolk, VA, vol 3, pp.1068–1072

Macwan A, Mosleh A (1994) A methodology for modeling operator errors of commission in probabilistic risk assessment. Reliabil Eng Syst Saf 45:139–157

Maurino DE, Reason J, Johnston N, Lee RB (1995) Beyond Aviation Human Factors – Safety in High Technology System. Ashgate, Aldershot, UK

Niwa Y, Hollnagel E (2002) Integrated computerization of operating procedures. Nuclear Eng Des 213:289–301

Niwa Y, Hollnagel E, Green M (1996) Guidelines for computerized presentation of emergency operating procedures. Nuclear Eng Des 167:113–127

O'Hara JM, Hall RE (1992) Advanced control rooms and crew performance issues: implications for human reliability. IEEE Trans Nuclear Sci 39(4):919–923

O'Hara JM, Higgins JC, Stubler WF, Kramer J (2002) Computer-based procedure systems: tech-

nical basis and human factors review guidance. NUREG/CR-6634, Washington, DC
Pirus D, Chambon Y (1997) The computerized procedures for the French N4 series. In: Proceedings of the 6th IEEE Conference on Human Factors and Power Plants, 3-9 June 1997, Orlando, FL.
Rasmussen J, Jensen A (1974) Mental procedures in real-life tasks: a case study of electronic trouble shooting. Ergonomics 17(3):293–307
Reason J (2008) The human Contribution: Unsafe Acts, Accidents and Heroic Recoveries. Ashgate, London
Reason J, Parker D, Lawton R (1998) Organizational controls and safety: The varieties of rule-related behavior. J Occupat Organizat Psychol 71:289–304
Reynes L, Beltranda GA (1990) Computerized control room to improve nuclear power plant operation and safety. Nuclear Saf 31(4):504–513
Spurgin AJ, Orvis DD, Cain DG, Yau CC (1988) Testing an expert system: Testing the emergency operating procedures tracking system. In: Proceedings of 4th IEEE Conference on Human Factors and Power Plants, Monterey, CA, pp.137–140
Spurgin AJ, Wachtel J, Moieni P (1993) The state of practice of computerized operating procedures in the commercial nuclear power industry. In: Proceedings on Human Factors and Ergonomics Society (HFES) Annual Meeting, Santa Monica, CA, pp.1014–1017
Stanton N (1996) Simulators: a review of research and practice. In: Stanton N (ed) Human Factors in Nuclear Safety, Taylor and Francis, London, pp.117–140
Sträter O, Bubb H (1999) Assessment of human reliability based on evaluation of plant experience: requirements and implementation. Reliabil Eng Syst Saf 63:199–219
Swain AD, Guttmann HE (1983) Handbook of human reliability analysis with emphasis on nuclear power plant applications. NUREG/CR-1278, Washington DC
Vessey I (1994) The effect of information presentation on decision making: a cost-benefit analysis. Inf Manage 27:103–119
Vicente KJ (2002) Ecological interface design: progress and challenges. Hum Factors 44(1):62–78
Vicente KJ (1999) Cognitive Work Analysis: Toward Safe, Productive and Healthy Computer-based Work. Erlbaum, Mahwah, NJ
Vicente KJ, Rasmussen J (1990) The ecology of human-machine systems II: Mediating direct perception in complex work domains. Ecol Psychol 2(3):207–249
Vicente KJ, Rasmussen J (1992) Ecological interface design: theoretical foundations. IEEE Trans Syst Man Cybern 22(4):589–606
Visciola M, Armando A, Bagnara S (1992) Communication patterns and errors in flight simulation. Reliabil Eng Syst Saf 36:253–259
Wickens CD (1992) Engineering Psychology and Human Performance, 2nd edn. Harper Collins, New York
Williams JC (1988) A data-based method for assessing and reducing human error to improve operational performance. In: Proceedings of the 4th IEEE Conference on Human Factors in Power Plants, Monterey, CA, pp.436–450
Woods DD (1991) The cognitive engineering of problem representations. In: Weir G, Alty JL (eds) Human-computer Interaction and Complex Systems, Academic, New York, pp.169–188
Woods DD, Rumancik JA, Hitchler MJ (1984) Issues in cognitive reliability. In: Lassahn PL, Majumdar D, Brockett GF (eds) Anticipated and Abnormal Plant Transients in Light Water Reactors, vol 2, Plenum, New York, pp.1127–1140,

11 Concluding Remarks with Outlook

Up to this point, a systematic framework called the TACOM measure, which can quantify the complexity of proceduralized tasks, has been explained from the beginning to validation. Actually, the results of validation activities show that there is a significant relation between TACOM scores and the performance of qualified operators.

Accordingly, we are able to say that the TACOM measure seems to be useful for quantifying the complexity of proceduralized tasks. Particularly, since TACOM scores could be used to identify complicated proceduralized tasks demanding an excessive workload on qualified operators, it is expected that the TACOM measure should be capable of providing an important clue for many pending issues.

11.1 Outlook for the TACOM Measure

In order to consider the outlook of the TACOM measure, comparing the applicable area of task analysis with that of the TACOM measure could provide valuable insights. Kirwan (1994) pointed out that the result of a task analysis will provide invaluable information supporting various areas, such as (1) allocation of function, (2) person specification, (3) interface design, (4) training procedures, (5) HRA, and (6) staffing and organization. For example, the results of the task analysis play an important role for extracting interface design specifications (i.e., *what controls/displays are necessary*?). Here, it seems that these areas are directly comparable to the promising applications of the TACOM measure because, as shown in Fig. 10.7, the results of a task analysis provide all kinds of necessary inputs for quantifying the five submeasures. This strongly implies that the applicable area of the TACOM measure can be extended as illustrated in Fig. 11.1.

For example, let us recall *verify pressurizer pressure is abnormally decreasing* action. As explained in Sect. 6.2.2, it is anticipated that qualified operators will be faced with a tricky decision (such as *which tendency represents abnormally decreasing pressurizer pressure*?) because the property of the *ACCEPTANCE CRITERION* is *SUB*. In this case, if qualified operators are not sufficiently trained, their responses will be diverse with respect to the situation at hand. In this regard, Leplat (1998) pointed out the following:

> Where this change can take place at the same level of processing, the same type of cognitive instruments are used, but in a different way. These different activities, which may be used in the execution of the same task, are often referred to as vicariants. The possibilities of vicariance are much greater when the task is loosely prescribed (the extreme case being a task where only the goal is prescribed) (p. 110).

Therefore, it is necessary to consider an additional training strategy by which qualified operators can be recognize how to cope with *the loosely prescribed tasks*.

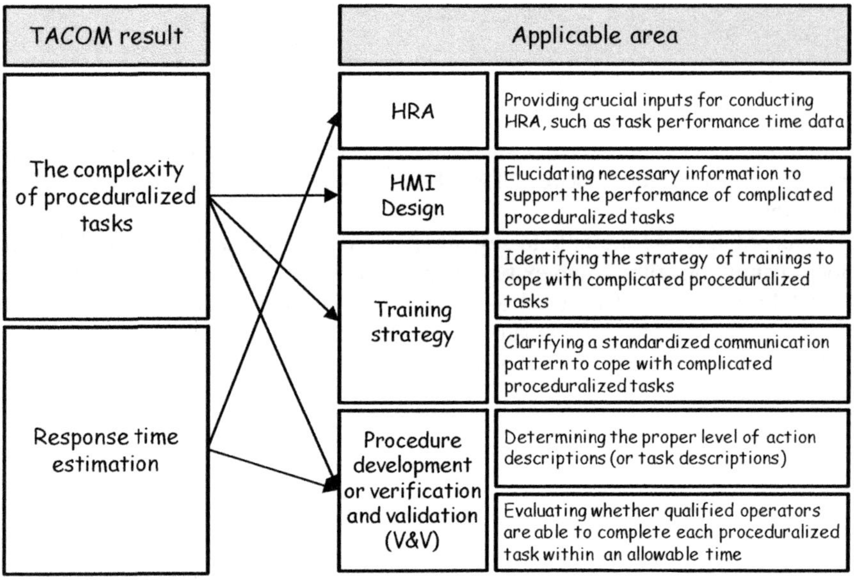

Fig. 11.1 Applicable area of the TACOM measure

In addition, the ambiguity of an action description could result in a communication problem. That is, when SROs are faced with this action, most of them are likely to give a command such as "RO, check whether the trend of pressurizer pressure is abnormally decreasing or not." In this case, if ROs just inform SROs of the observable tendency of pressurizer pressure (i.e., "The trend recorder says that pressurizer pressure is decreasing now") without any notification, then SROs will likely decide the pressurizer pressure is abnormally decreasing. This means that a standardized communication protocol that allows qualified operators to correctly convey what they are concerned about should be emphasized in the course of training.

Moreover, it is meaningful to scrutinize the effect of task complexities on changes in communication patterns because it is believed that the possibility of inappropriate communications would increase in proportion to an increase in workload. For example, Urban et al. (1996) reported that team members decreased the amount of communications when the workload increased. In addition, it was observed that qualified operators frequently change their communication patterns in order to cope with a decrease in available time (Kontogiannis 1999b) or to accomplish a task demanding a long task performance time (Visciola et al. 1992).

However, a more interesting application of the TACOM measure would be the provision of an insightful clue to determine the appropriate level of an action description (or a task description) because this is one of crucial pending issues in procedure development (DOE 1998; Inaba et al. 2004; Wieringa et al. 1998). For example, let us consider the following two actions adopted from Zach (1980):

- Isolate letdown line
- Isolate letdown line by closing valves CV1214 and CV1216

Regarding these actions, it has been reported that most SROs (i.e., highly experienced qualified operators) preferred the former, while less experienced qualified operators (such as ROs and TOs) preferred the latter. Here, it should be emphasized that there is a clear difference in the action descriptions. That is, there is no specification about *MEANS* in the former (i.e., *NM*), while the latter has an obvious specification about *MEANS* (i.e., *DEG*). This indicates that the description level of the former is lower than that of the latter.

However, the problem is that we need to establish a firm standard allowing us to consistently describe an action, since a good procedure should provide crucial contents with which even *less experienced qualified operators* can properly perform the required actions in a real situation. From this concern, it is evident that the level of action descriptions should be determined by a combination of the properties of three radical elements: *MEANS*, *ACCEPTANCE CRITERION*, *CONSTRAINT*, and with a *peculiarity*. Consequently, if we elucidate a relationship between the preference of qualified operators and the characteristics of action specifications, then we could develop practical guidelines that are serviceable to determine the proper level of action descriptions.

It is still true that we have to devote huge amounts of additional effort to resolving practical problems pertaining to the TACOM measure. For example, improvement of the TACOM measure is indispensable, because the TACOM measure has intrinsic limitations, such as a lack of ability to consider the effect of a task environment as well as the effect of personality on the complexity of proceduralized tasks (Fig. 3.7).

In addition, it is necessary to reduce the difficulty calculating TACOM scores. As explained in Chap. 7, the TACOM score of each proceduralized task can be calculated by following eight processes. Unfortunately, since these processes are somewhat tricky, the analysis of procedures that consist of many proceduralized tasks is probably more difficult than it seems. It is to be noted that, to resolve this problem, a TACOM calculator that provides a graphical user interface facilitating the quantification of five kinds of submeasures is now available (*Appendix C*).

Nevertheless, according to research activities and the associated results presented throughout this book, we are able to suggest a new research area tentatively called *cognitive procedure engineering* (CPE), by which practical as well as effective solutions can be deduced to minimize the amount of undue workload felt by qualified operators. In other words, in contrast to traditional approach that largely deals with physical characteristics from an ergonomics or human factors perspective (e.g., focusing on sentence structures, font sizes, writing styles, vocabularies, etc.), it is believed that the TACOM measure will be a trailblazer in the development of an *engineered* procedure that considers the cognitive characteristics of qualified operators. Based on this belief, I would like to end my book by drawing a simple but decisive conclusion as follows.

Since the TACOM measure seems to properly quantify the complexity of proceduralized tasks, it is highly expected that insightful clues for resolving many pending issues related to developing a good procedure can be obtained from a novel viewpoint considering the cognitive characteristics of qualified operators.

References

DOE (1998) Writer's guide for technical procedures. DOE-STD-1029-92

Inaba K, Parsons SO, Smillie R (2004) Guidelines for Developing Instructions. CRC, Boca Raton, FL

Kirwan B (1994) A Guide to Practical Human Reliability Assessment. Taylor and Francis, London

Kontogiannis T (1999b) Training effective human performance in the management of stressful emergencies. Cognit Technol Work 1:7–24

Leplat J (1998) Task complexity in work situations. In: Goodstein LP, Anderson HB, Olsen SE (eds) Tasks, Errors and Mental Models, Taylor and Francis, London, pp.105–115

Urban JM, Weaver JL, Bowers CA, Rhodenizer L (1996) Effects of workload and structure on team processes and performance: implications for complex team decision making. Hum Factors 38(2):300–310

Visciola M, Armando A, Bagnara S (1992) Communication patterns and errors in flight simulation. Reliabil Eng Syst Saf 36:253–259

Wieringa D, Moore C, Barnes V (1998) Procedure Writing Principles and Practices, 2nd edn. Battelle Press, Columbus, OH

Zach SE (1980) Control room operating procedures: content and format. In: Proceedings of the 24th Annual Meeting of the Human Factors and Ergonomics Society (HFES), Los Angeles, CA, pp.125–127

Part IV
Appendices

Appendix A Categories of Complexity Factors

A1 Amount of Information

Reference	Complexity factor specified in reference
Benbasat and Taylor (1982)	Factors resulting in information load: • Number of dimensions extracted from data • Fitness of discrimination process • Number of interconnections among rules for combining data
Bui and Sivasankaran (1990)	Amount of data
Byström and Järvelin (1995)	Information load
Campbell (1988)	Information load
Jacko and Salvendy (1996)	Number of cues
Leplat (1998)	Size of memory set
Li and Wieringa (2000)	Amount of information to maintain in working memory
Maynard and Hakel (1997)	Information load
Roth et al. (1992)	Amount of information
Stassen et al. (1990)	Information load
Sundstrom (1993)	Number of indications associated with operational states
Svensson et al. (1997)	Number of symptoms
Thelwell (1994)	Number of alarms and symptoms
Visser and Wieringa (2001)	Number of alarms
Wei et al. (1998)	Number of stimuli
Wood (1986)	Number of information cues to be processed in performance of each act
Wood and Locke (1990)	Number of information cues

A2 Number of Actions

Reference	Complexity factor specified in reference
Chi and Chung (1996)	Total number of elementary task unit
Jacko and Salvendy (1996)	Number of commands necessary
Kieras and Polson (1985)	Number of operators (physical activities)
Leplat (1998)	Number of elements or units
Li and Wieringa (2000)	• Number of steps to be performed for achieving a task • Number of tasks
Schmuck and Gundlach (1989)	Number of cognitive steps (i.e., the number of cognitive activities)
Sundstrom (1993)	Number of steps required to reach an desired goal
Thelwell (1994)	Number of actions
Wei et al. (1998)	Number of required actions
Wood (1986)	• Number of subtasks • Number of distinct acts in a subtask
Wood and Locke (1990)	Number of acts required to complete task

A3 Logical Entanglement

Reference	Complexity factor specified in reference
Campbell (1988)	Multiple path-goal connections
Jacko and Salvendy (1996)	Path-goal multiplicity
Kieras and Polson (1985)	Number of methods (i.e., execution sequences to achieve a goal)
Leplat (1998)	Size of acquisition hierarchy required for task execution
Li and Wieringa (2000)	Links and dependencies among tasks
Rouse and Rouse (1979)	Number of relevant relationships to available symptoms
Sundstrom (1993)	Interrelatedness of required steps
Thelwell (1994)	Relationship between actions and events
Wood (1986)	Number of precedence relations among distinct acts
Wood and Locke (1990)	Sequencing of acts required to complete a task

A4 Amount of Domain Knowledge

Reference	Complexity factor specified in reference
Allen et al. (1996)	• Number of components in system • Number of relevant relationships between components
Li and Wieringa (2000)	Amount of knowledge to extract from long-term memory
Morris and Rouse (1985)	• Number of components included in network • Number of relevant relationships between components
Leplat (1998)	• Number of elements or units • Relations among elements or units
Liao and Palvia (2000)	• Number of objects • Degree of relationships between objects • Degree of nesting of objects • Number of generalization hierarchies
Rouse (1978)	Problem size (i.e., number of components included in network)
Rouse and Rouse (1979)	Number of components

A5 Level of Engineering Decision

Reference	Complexity factor specified in reference
Kieras and Polson (1985)	• Number of operators (cognitive activities) • Number of selection rules (i.e., number of decisions to select an appropriate method)
Schmuck and Gundlach (1989)	Number of cognitive steps (i.e., number of cognitive activities)
Sundstrom (1993)	• Interrelatedness of assessment, choice and evaluation rules • Interconnectedness of operational states • Relation between indicators and operational states • Number of assessments, choices and evaluation rules • Number and relation between conditions for assessments, choices and evaluation rules
Svensson et al. (1997)	Number of decisions
Thelwell (1994)	Number of decisions

A6 Time Pressure

Reference	Complexity factor specified in reference
Allen et al. (1996)	Time constraints
Thelwell (1994)	Time available
Morris and Rouse (1985)	Time constraints
Svensson et al. (1997)	Time pressure
Umbers (1979)	Time pressure
Leplat (1998)	Time pressure
Payne et al. (1988)	Time pressure
Bui and Sivasankaran (1990)	Time pressure
Rouse (1978)	Time pressure
Hirotsu et al. (2001)	Time pressure
Wei et al. (1998)	Time pressure
Stassen et al. (1990)	Time pressure
Wood and Locke (1990)	Time allowed for performance of a task

A7 Temporal Characteristics

Reference	Complexity factor specified in reference
Decortis (1993)	• Event frequency • Chronology of events
Leplat (1998)	Temporal override of task currently being performed
Li and Wieringa (2000)	• Nature and diversity of task. • Uncertainty of arrival rate of occurrence and duration of tasks
Thelwell (1994)	• Number of malfunctions • Rate of appearance of new tasks • Sequencing and frequency with which activity/event occurs
Wei et al. (1998)	Degree of overlap of multiple task demands

A8 System Characteristics

Reference	Complexity factor specified in reference
Leplat (1998)	Delayed nature of feedback Redundancy of a stimulus ensemble
Sundstrom (1993)	Dynamicity of technical system Indicator variability (rate of change)
Roth, Mumaw and Stubler (1992)	Difficulty in accessing required information

A9 Personal Characteristics

Reference	Complexity factor specified in reference
Li and Wieringa (2000)	• Intelligence • Personality • Cultural background • Willingness
Maynard and Hakel (1997)	• Cognitive ability • Task motivation
Morris and Rouse (1985)	• Abilities (aptitudes) • Cognitive style
Rouse and Rouse (1982)	• Human ability • Aptitudes • Cognitive style

Appendix B Task Performance Time Data Obtained from Reference NPPs

ID	SIC	SLC	SSC	AHC	EDC	Avg. (s)[1]	SD (s)[2]
1	2.322	1.585	0.918	1.922	1.922	3.4	0.9
2	2.000	1.585	0.918	2.585	1.922	11.2	7.3
3	2.807	1.585	0.918	1.922	2.585	9.4	6.5
4	1.922	1.500	2.000	3.170	2.128	15.5	11.8
5	3.170	1.500	2.000	2.128	2.128	15.1	9.9
6	3.170	1.500	2.000	2.128	2.128	20.4	21.2
7	3.000	1.585	0.918	2.585	2.807	27.5	15.0
8	3.322	1.922	2.322	2.281	2.281	27.4	19.1
9	3.585	1.371	2.322	2.281	2.281	21.7	12.5
10	2.585	1.500	2.000	3.170	2.750	9.3	4.5
11	2.585	2.000	2.000	3.170	2.750	7.8	4.8
12	3.585	1.922	2.322	2.281	2.281	29.0	20.1
13	3.665	1.252	2.585	2.404	2.404	37.2	21.3
14	3.700	1.252	2.585	2.404	2.404	21.0	6.1
15	2.322	1.500	2.000	3.170	3.459	18.5	7.4
16	3.170	1.922	2.322	3.122	2.281	10.8	5.4
17	3.459	1.922	2.322	2.846	2.281	29.8	20.4
18	3.700	1.252	2.585	2.918	2.404	16.0	9.6
19	3.000	1.371	2.322	3.585	2.846	9.8	5.8
20	4.000	1.793	2.585	2.404	2.404	37.2	25.8
21	3.322	1.793	2.585	2.918	3.085	14.8	5.2
22	4.000	1.665	2.807	2.507	2.507	40.7	32.3
23	2.750	2.322	2.322	3.585	2.846	15.5	9.2
24	2.322	1.922	2.322	3.585	3.585	9.0	4.6
25	3.585	1.371	2.322	3.585	2.846	19.0	12.7
26	3.000	1.922	1.922	3.278	4.000	20.0	20.3
27	0.000	1.793	2.585	4.000	4.170	29.7	18.7

28	3.459	2.322	2.322	3.585	2.846	19.5	12.5
29	3.278	1.842	2.807	3.970	2.507	29.5	22.3
30	4.322	1.549	3.000	3.031	2.597	38.3	16.5
31	2.807	1.918	2.585	3.970	3.774	17.0	6.6
32	4.170	1.665	2.807	3.133	3.236	33.2	21.1
33	4.170	1.842	2.807	2.9781	3.3736	20.5	13.3
34	3.322	1.149	2.807	4.170	3.374	30.0	15.4
35	4.755	1.447	3.170	2.676	2.676	32.2	9.8
36	3.700	1.549	3.000	3.405	3.525	43.3	17.0
37	3.665	1.252	2.585	3.907	3.590	27.7	15.4
38	4.524	2.059	3.170	3.432	2.676	39.1	13.7
39	4.088	1.149	2.807	4.170	3.374	21.7	16.1
40	3.700	1.842	2.522	3.703	4.297	30.9	23.8
41	3.700	1.842	2.522	3.703	4.297	23.0	18.5
42	4.644	1.880	3.170	3.078	3.432	58.8	35.1
43	3.907	1.252	2.585	4.000	4.248	35.6	3.7
44	4.143	1.149	2.807	4.170	3.970	42.2	14.4
45	3.468	1.658	3.170	4.502	3.849	43.6	21.5
46	4.000	1.793	2.585	4.000	4.392	30.0	13.1
47	4.840	1.559	3.907	3.829	2.856	52.6	25.3
48	4.088	1.665	2.807	4.059	4.564	41.0	19.7
49	4.533	1.793	3.418	3.940	3.741	57.0	33.9
50	4.533	1.793	3.418	3.940	3.741	47.9	23.7
51	4.533	1.793	3.418	3.940	3.741	44.1	16.7
52	4.890	1.371	3.323	3.763	3.763	64.8	13.8
53	4.143	1.061	3.000	4.459	4.297	55.3	32.4
54	4.248	1.149	2.807	4.248	4.524	51.9	16.5
55	4.369	1.549	3.000	4.392	3.998	54.7	22.7
56	5.333	1.549	3.875	3.390	3.390	83.9	39.7
57	4.459	1.959	3.418	4.220	3.970	61.9	43.1
58	4.907	1.278	3.459	4.638	3.163	58.0	37.7
59	3.808	1.722	3.322	4.907	4.750	26.6	17.3
60	3.684	1.921	3.665	4.811	4.631	77.3	88.3
61	4.585	1.892	3.547	4.558	4.345	63.2	47.9
62	4.585	1.585	3.585	4.75	4.323	44.7	43
63	3.170	1.278	3.459	5.1293	5.459	48.0	24.4
64	4.222	1.868	3.459	4.789	4.901	32.0	11.1
65	4.907	2.264	3.522	4.113	4.736	72.2	30.4

66	5.426	1.145	3.700	3.237	5.170	69.3	32.2
67	5.322	1.145	3.700	3.346	5.210	60.6	24.2
68	4.954	1.145	3.700	5.1554	4.107	78.3	37.8
69	5.380	2.032	4.166	4.407	3.750	130.3	57.5
70	4.863	1.149	3.236	5.047	4.871	152.5	52.3
71	5.114	1.769	3.837	5.297	3.814	85.4	35.6
72	5.114	1.769	3.837	5.297	3.814	37.1	29.3
73	5.114	1.769	3.837	5.297	3.814	89.0	62.3
74	4.392	2.138	4.524	5.003	4.549	62.3	19.5
75	3.807	1.549	3.625	5.512	5.772	90.7	40.5
76	5.072	2.259	3.641	5.052	5.132	155.6	109.2
77	4.907	2.105	4.170	5.272	4.884	47.3	17.5
78	4.897	2.173	4.316	5.414	5.223	71.2	20.1
79	5.802	1.942	4.430	5.330	5.204	196.5	36.8
80	5.728	1.515	4.236	6.051	5.206	139.3	46.1
81	5.736	1.987	4.228	5.645	5.524	264.3	80.7
82	5.961	2.276	3.992	5.752	5.931	159.9	54.0
83	6.327	2.248	4.449	5.670	5.336	275.5	119.6
84	5.132	1.945	4.260	6.318	6.431	200.1	47.6
85	5.722	2.125	4.595	6.121	5.960	183.7	41.0
86	5.668	2.113	4.761	6.121	6.214	169.0	66.7
87	6.329	1.873	4.682	5.655	6.178	507.0	239.4
88	5.544	1.988	4.396	6.346	6.458	182.5	115.0
89	5.638	2.037	4.594	6.413	6.420	226.0	263.8
90	6.584	1.702	4.320	6.144	6.226	280.3	176.2
91	5.926	1.624	5.088	6.200	6.591	122.4	33.6

[1]Avg. denotes the averaged task performance time measured in seconds.
[2]SD: standard deviation measured in seconds.

Appendix C Brief Introduction to the TACOM Calculator

As outlined in Chap. 7, we have to complete eight phases to get the values of the five submeasures by which the TACOM score of a proceduralized task can be calculated. Unfortunately, it is quite tricky to carry out each phase with bare hands. For example, the identification of DAs, which is the main purpose of the third phase, is very laborious. This is because not only a *peculiarity* but also every property of the three radical elements about action specifications (i.e., *MEANS*, *ACCEPTANCE CRITERION*, and *CONSTRAINT*) should be compared for all the required actions included in an action analysis form (Table 7.3). The identification of DI has a similar problem. The construction of necessary graphs (such as ACGs, ISGs, AHGs, and EDGs) would be another source of difficulty in quantifying the value of each submeasure.

Accordingly, dedicated software called the TACOM calculator (version 1.0) has been developed. The architecture of the TACOM calculator was designed along with well-known guidelines, and necessary activities pertaining to the quality assurance (QA) of the TACOM calculator were also performed (IEEE 2000; ISO 1991; USNRC 1993). The following points summarize the system requirement of the TACOM calculator:

- Hardware: IBM-compatible Pentium-based PC, Pentium 4 or later CPU, more than 512MB system memory, at least 122 MB available hard disk space for system and 470 MB for database
- Operating system: Windows 2000 or later
- Database: MySQL version 5.0 for Windows X86

Figure C.1 shows the initiation image of the TACOM calculator, which consists of five panes with distinctive functions. In addition, Fig. C.2 shows an example related to the quantification of the five submeasures about *increasing the rate of charging flow* task (Fig. 10.7).

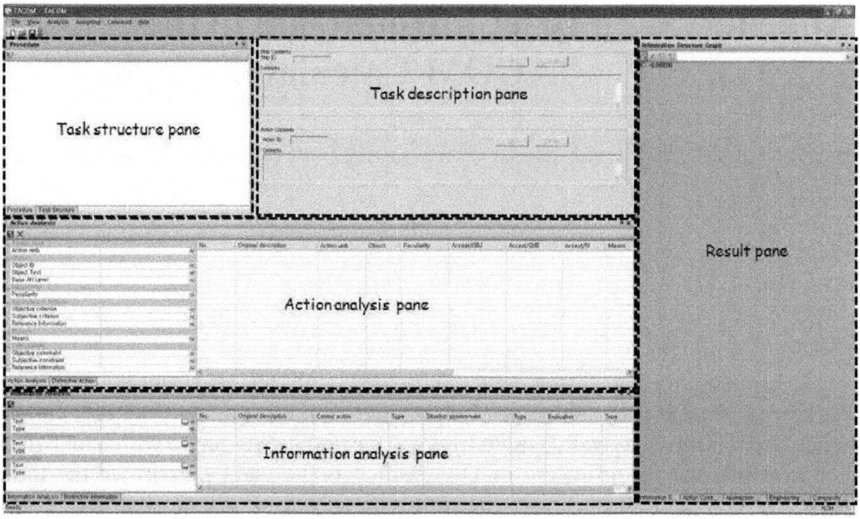

Fig. C.1 Iitiation image of the TACOM calculator

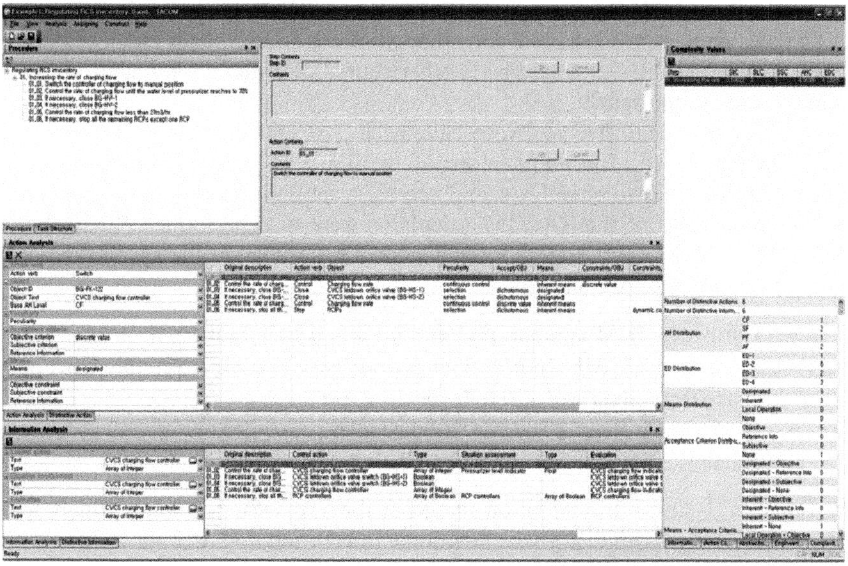

Fig. C.2 Quantifying the complexity of *increasing the rate of charging flow* task

First, a task structure should be defined. Since the task being considered consists of six actions, it is necessary to add each action to the *task structure pane*. After that, for each action, an original action description should be given in the *task description pane*. Figure C.3 depicts an example of the definition as well as the original description about *switch the controller of charging flow to manual position* action.

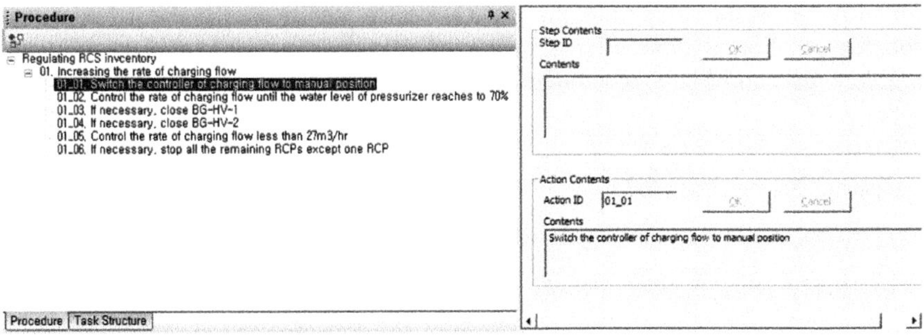

Fig. C.3 Defining a task structure with the associated actions

When a task structure is defined with the associated actions, it is necessary to clarify a *peculiarity* as well as the properties of the three radical elements about action specifications in the *action analysis pane*, which corresponds to an action analysis form. Similarly, it is necessary to clarify the source of information to be processed by qualified operators in the *information analysis pane* (i.e., information analysis form). Figures C.4a and C.4b show an example of filling out the *action analysis pane* and the *information analysis pane* with respect to the task being considered, respectively.

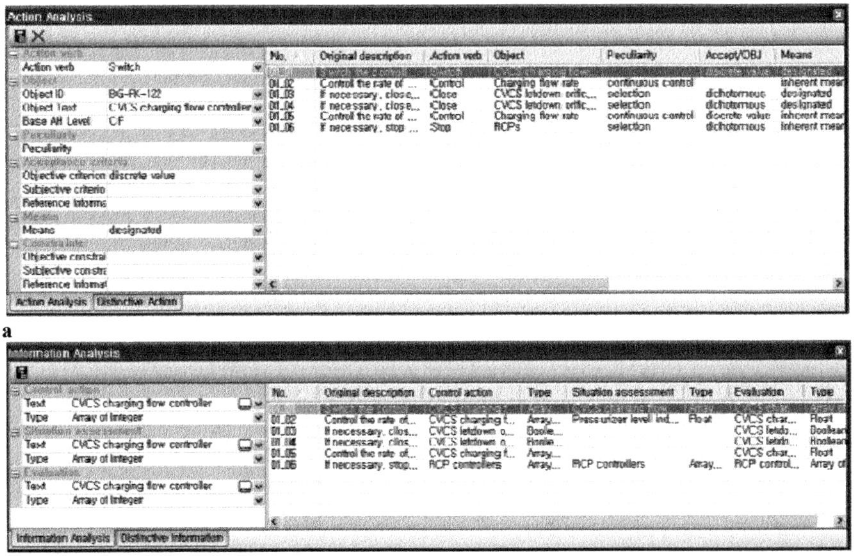

a

b

Fig. C.4 An example of filling out the *action analysis pane* and the *information analysis pane*

If all the necessary inputs are properly provided, the TACOM calculator is able to identify the list of DAs and DI. In addition, based on these lists, the TACOM calculator automatically generates the associated graphs in the *result pane*, except for an ACG. At the same time, the TACOM calculator quantifies the value of each submeasure. For example, Fig. C.5 shows the ISG of *increasing the rate of charging flow* task, and Fig. C.6 summarizes the results of quantifying the five submeasures.

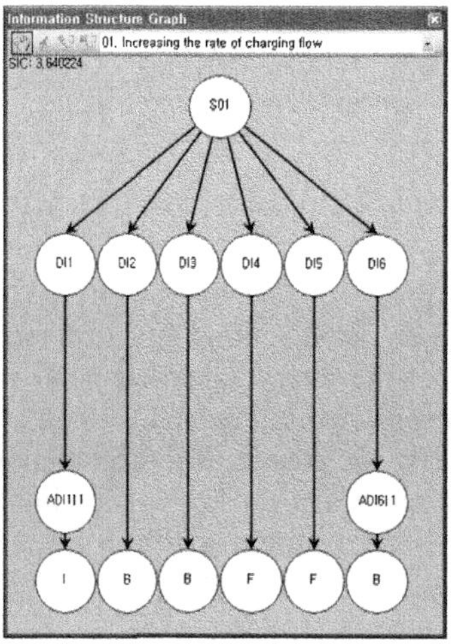

Fig. C.5 An example ISG of *increasing the rate of charging flow* task, which is automatically generated by the TACOM calculator

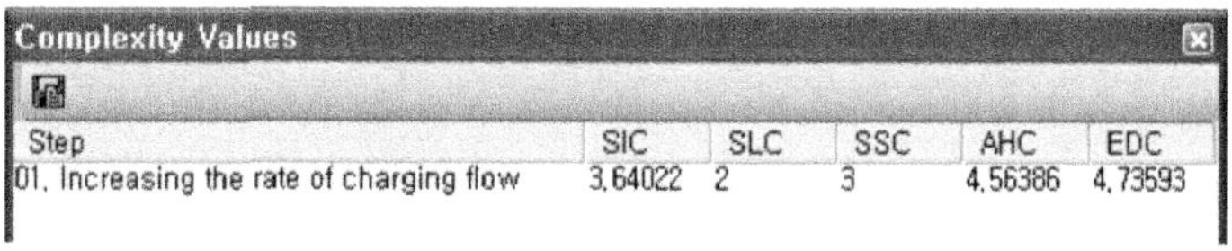

Complexity Values

Step	SIC	SLC	SSC	AHC	EDC
01. Increasing the rate of charging flow	3.64022	2	3	4.56386	4.73593

Fig. C.6 The value of each submeasure pertaining to *increasing the rate of charging flow* task

Please inquire at the following address to ontain more information about the TACOM calculator.

TACOM administrator

Integrated Safety Assessment Division
Korea Atomic Energy Research Institute
1045 Daedeokdaero, Yuseong-Gu, Daejeon, 305-353, Korea
Fax: +82-42-868-8256
E-mail: tacomadmin@kaeri.re.kr

References Appearing in Appendices

Allen JA, Teague RC, Carter RE (1996) The effects of network size and fault intermittency on troubleshooting performance. IEEE Trans Syst Man Cybern A Syst Hum 26(1):125–132

Benbasat I, Taylor RN (1982) Behavioral aspects of information processing for the design of management information systems. IEEE Trans Syst Man Cybern 12(4):439–450

Bui T, Sivasankaran TR (1990) Relation between GDSS use and group task complexity: an experimental study. In: Proceedings of the 23rd Annual Hawaii International Conference on System Science, Kailua-Kona, vol 3, pp.69-78

Byström K, Järvelin K (1995) Task complexity affects information seeking and use. Inf Process Manage 31(2):191-213

Campbell DJ (1988) Task complexity: a review and analysis. Acad Manage Rev 13(1):40-52

Chi C, Chung K (1996) Task analysis for computer-aided design (CAD) at a keystroke level. Appl Ergonom 27(4): 255-265

Decortis F (1993) Operator strategies in a dynamic environment in relation to an operator model. Ergonomics 36(11):1291-1304

Hirotsu Y, Suzuki K, Kojima M, Takano K (2001) Multivariate analysis of human error incidents occurring at nuclear power plants: several occurrence patterns of observed human errors. Cogn Technol Work 3:82-91

IEEE (2000) Recommended practice for architectural description of software-intensive system. IEEE Std 1471-2000, Piscataway, NJ

ISO (1991) Information technology: software product evaluation-quality characteristics and guidelines for their use. ISO/IEC 9126, Geneva

Jacko JA, Salvendy G (1996) Hierarchical menu design: breath, depth and task complexity. Percept Mot Skills 82:1187-1201

Kieras D, Polson PG (1985) An approach to the formal analysis of user complexity. Int J Man-Mach Stud 22:365-394

Leplat J (1998) Task complexity in work situations. In: Goodstein LP, Anderson HB, Olsen SE (eds) Tasks, Errors and Mental Models, Taylor and Francis, London, pp.105-115

Li K, Wieringa PA (2000) Understanding perceived complexity in human supervisory control. Cogn Technol Work 2:75-88

Liao C, Palvia PC (2000) The impact of data models and task complexity on end-user performance: an experimental investigation. Int J Hum-Comput Stud 52:831-845

Maynard DC, Hakel MD (1997) Effects of objective and subjective task complexity on performance. Hum Perform 10(4):303-330

Morris NM, Rouse WB (1985) Review and evaluation of empirical research in troubleshooting. Hum Factors 27(5):503-530

Payne JW, Bettman JR, Johnson EJ (1988) Adaptive strategy selection in decision making. J Exp Psychol: Learn, Mem Cognit 14(3):534-552

Roth EM, Mumaw RJ, Stubler WF (1992) Human factors evaluation issues for advanced control rooms: a research agenda. In: Proceeding of the 5th IEEE Conference on Human Factors and Power Plants, Monterey, CA, pp.254-259

Rouse SH, Rouse WB (1982) Cognitive style as a correlate of human problem solving perfor-

mance in fault diagnosis tasks. IEEE Trans Syst Man Cybern 12(5):649-652
Rouse WB (1978) Human problem solving performance in a fault diagnosis task. IEEE Trans Syst Man Cybern 8(4):258-271
Rouse WB, Rouse SH (1979) Measures of complexity of fault diagnosis task. IEEE Trans Syst Man Cybern 9(11):720-727
Schmuck P Gundlach W (1989) Reduction of mental effort in tasks of different complexity. In: Klix F, Streitz NA, Waern Y, Wandke H (eds) Man-computer Interaction Research. Elsevier, Amsterdam
Stassen HG, Johannsen G, Moray N (1990) Internal representation, internal model, human performance model and mental workload. Automatica 26(4):811-820
Sundstrom GA (1993) Towards models of tasks and task complexity in supervisory control applications. Ergonomics 36:1413-1423
Svensson E, Angelbrog-Thandrez M, Sjoberg L, Olsson S (1997) Information complexity – mental workload and performance in combat aircraft. Ergonom 40:362-380
Thelwell PJ (1994) What defines complexity? In: Robertson SA (ed) Contemporary Ergonomics: Ergonomics for All, Taylor and Francis, London, pp.89-94
Umbers IG (1979) Models of the process operator. Int J Man-Mach Stud 11:263-284
USNRC (1993) Software quality assurance program and guidelines. NUREG/BR-0167, Washington DC
Visser M, Wieringa PA (2001) PREHEP: Human error probability based process unit selection. IEEE Trans Syst Man Cybern C Appl Rev 31(1):1-15
Wei ZG, Macwan AP, Wieringa PA (1998) A quantitative measure for degree of automation and its relation to system performance and mental load. Hum Factors 40(2):277-295
Wood RE (1986) Task complexity: definition of the construct. Organizat Behav Hum Decis Processes 37:60-82
Wood RE, Locke EA (1990) Goal setting and strategy effects on complex tasks. Res Organizat Behav 12:73-109

Index

P

Q

R

S

GPSR Compliance

The European Union's (EU) General Product Safety Regulation (GPSR) is a set of rules that requires consumer products to be safe and our obligations to ensure this.

If you have any concerns about our products, you can contact us on ProductSafety@springernature.com

In case Publisher is established outside the EU, the EU authorized representative is:

Springer Nature Customer Service Center GmbH
Europaplatz 3
69115 Heidelberg, Germany

Batch number: 10755053

Printed by Printforce, the Netherlands